The present studies of coronoid systems is a natural continua-
tion of the corresponding studies of benzenoid systems. Both topics
are rooted in organic chemistry through certain polycyclic conjugated
hydrocarbons, which are the chemical counterparts of the systems in
question. However, the scope of the present work and corresponding
works on benzenoids goes far beyond a chemical motivation. These works
are classified under mathematical chemistry, a relatively new designa-
tion.

The book is supposed to have an interest for organic chemists
within certain specialities, but still more for theoretical and mathematical chemists. The last category has been characterized as *enfants
terribles* in the foreword of the first issue of the *Journal of Mathematical Chemistry* (1987). Finally, this book may have a considerable
interest for mathematicians within combinatorics and graph theory.

It is supposed that the book will be most useful for researchers,
including graduate students, in the pertinent fields. The text contains
no advanced mathematics whatsoever and should as such not represent any
barrier even for undergraduate students.

Here we wish to make some comments on the terminology, which is
not standardized and is partly controversial in this field of topological studies of polyhex (benzenoid and coronoid) systems.

The figure on the next page, a preliminary survey of some classes
of coronoids, incidently exemplified by systems with 14 hexagons each,
should give a first idea of what it all is about. The term "coronoid"
(Brunvoll, Cyvin BN and Cyvin 1987a) was originally devised so as to
be, as far as possible, analogous with "benzenoid". As to the latter
concept (benzenoid) we adher to the definition which was adopted in a
previous volume of *Lecture Notes in Chemistry* (Cyvin and Gutman 1988).

It is seen from the preliminary survey that a coronoid in our
terminology may be either catacondensed or pericondensed. The same is
true for a benzenoid. In the recently proposed terminology by
Trinajstić (1990a) the coronoids are not allowed to be catacondensed
or pericondensed. On the other hand, in consistence with our terminology, they can be Kekuléan or non-Kekuléan. But also here the analogy

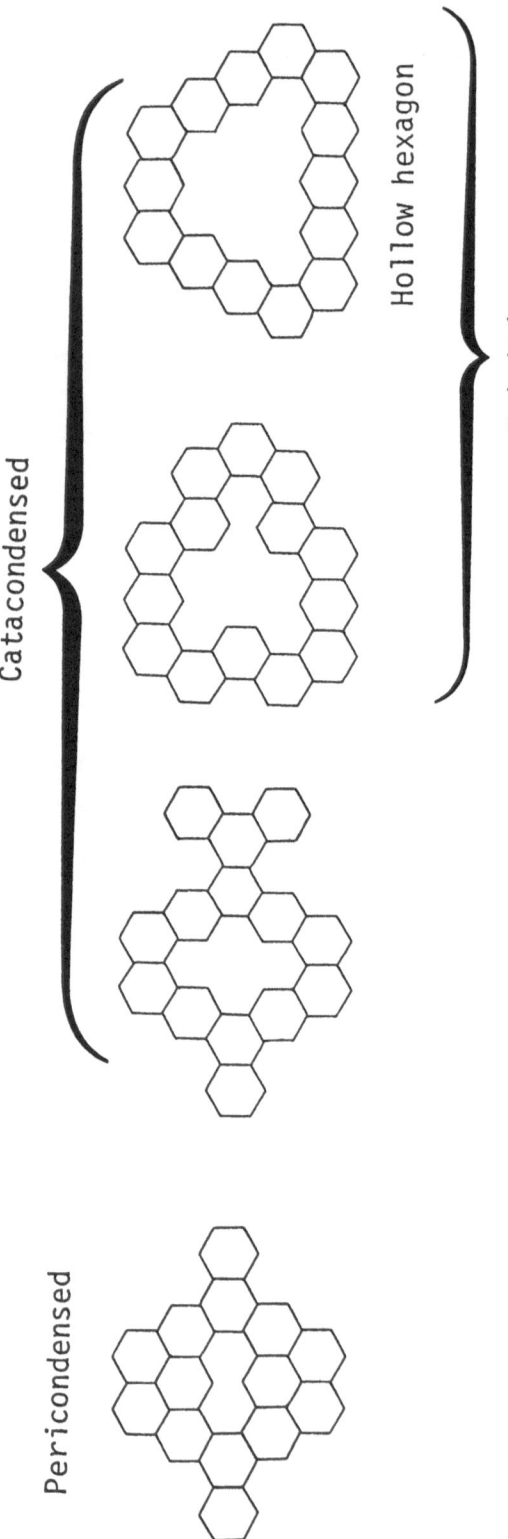

Periphery labels: Pericondensed, Catacondensed, Primitive, Hollow hexagon

Preliminary survey of some classes of single coronoids.

Lecture Notes in Chemistry

Edited by G. Berthier M.J.S. Dewar H. Fischer
K. Fukui G.G. Hall J. Hinze H.H. Jaffé J. Jortner
W. Kutzelnigg K. Ruedenberg J. Tomasi

54

S.J. Cyvin J. Brunvoll B.N. Cyvin

Theory of Coronoid Hydrocarbons

Springer-Verlag

Berlin Heidelberg New York London
Paris Tokyo Hong Kong Barcelona

Authors

S.J. Cyvin
J. Brunvoll
B.N. Cyvin
The University of Trondheim
The Norwegian Institute of Technology
Division of Physical Chemistry
N-7034 Trondheim - NTH Norway

ISBN 978-3-540-53577-5   ISBN 978-3-642-51110-3 (eBook)
DOI 10.1007/978-3-642-51110-3

2152/3140-543210 – Printed on acid-free paper

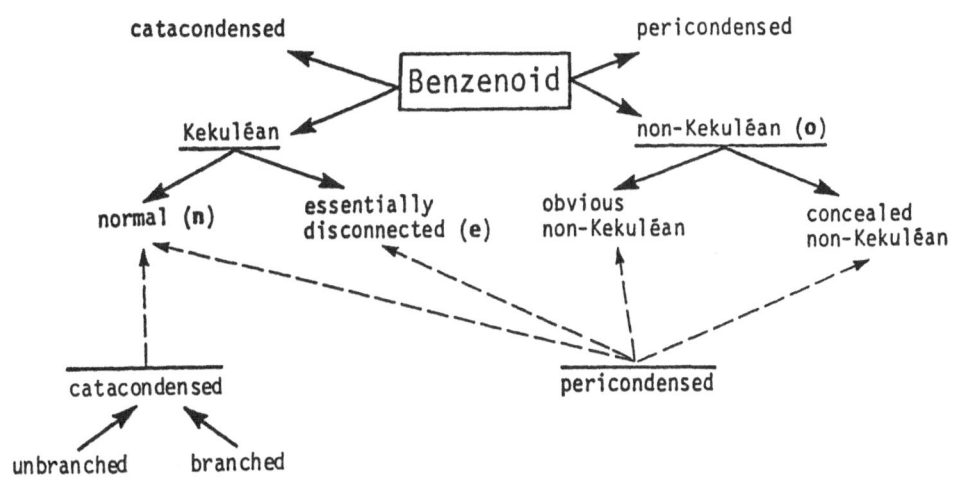

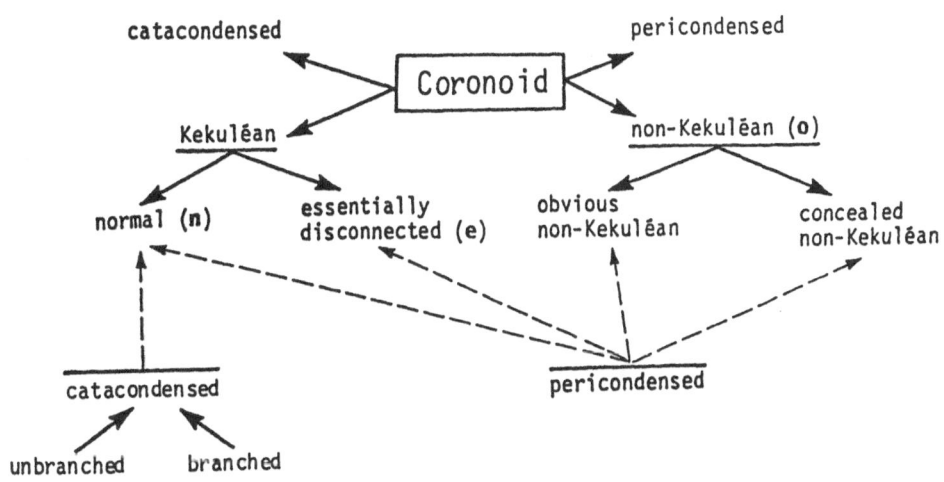

Idential classification schemes for benzenoids (top) and coronoids (bottom).

between benzenoids and coronoids stops in the scheme of Trinajstić (1990a), because he has reserved the term "benzenoid" for exclusively Kekuléan systems.

In the above figure we show a classification scheme for benzenoids, which is seen to be transferable in all details to coronoids.

Sven J. Cyvin

October 1990   Jon Brunvoll

Bjørg N. Cyvin

CONTENTS

PART I

INTRODUCTION

Chapter 1

CORONOID HYDROCARBONS

1.1 INTRODUCTION

 Coronoid systems, which are defined in the subsequent chapter, have chemical
counterparts in what we shall call *coronoid hydrocarbons*. Different other terms
have been introduced in organic chemistry, some of which also being used in the
present chapter. All the compounds to be considered here belong to the conjugated
hydrocarbons. Typical coronoid hydrocarbons are polycyclic (aromatic) and consist
of benzenoid rings. They are macrocyclic in contrast to the benzenoid hydrocarbons.

 The only few coronoid hydrocarbons which have been synthesized belong to a
class called *cycloarenes* (Staab and Diederich 1983). These compounds are cyclic
catacondensed benzenoid rings forming a macrocyclic system so that a cavity is present, into which carbon–hydrogen bonds are pointing. The idea should be clear from
Fig. 1. *Cycloarenes* have also been called "coronaphenes" (Jenny and Peter 1965a).

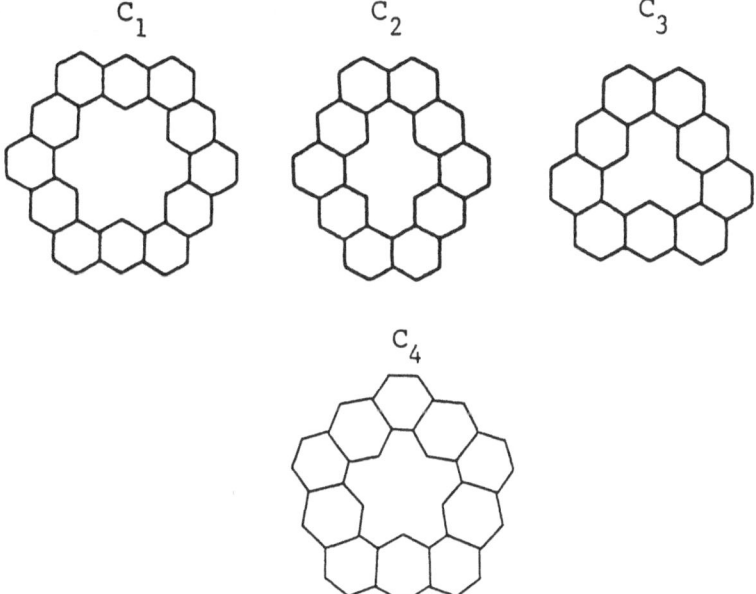

Fig. 1.1. Four *cycloarenes*: $C_{48}H_{24}$ (C_1), $C_{40}H_{20}$ (C_2), $C_{36}H_{18}$ (C_3) and $C_{40}H_{20}$ (C_4),
which is isomer with C_2.

In the following we shall also take into account hydrocarbons which do not correspond to coronoid systems according to the strict definition (see next chapter). The molecule C_4 in Fig. 1 is the first example. It is a matter of definition whether molecules like this should be designated coronoid hydrocarbons. We shall make allowance for such usage of the concept without detailed specifications. Also included in the present chapter are those classes of hydrocarbons of which some members correspond to what we shall call degenerate coronoid systems (and one quasi-coronoid system); precise definitions are given in Chapter 2.

1.2 CYCLOARENES

1.2.1 *The Story of Kekulene*

The route to the synthesis of a *cycloarene* has been long and troublesome (Peter and Jenny 1966; Vögtle and Staab 1968; see also the review: Staab and Diederich 1983). Here we shall report on the efforts which were concentrated around the synthesis of the $C_{48}H_{24}$ molecule (C_1 of Fig. 1) or *cyclo*[d.e.d.e.d.e.d.e.-d.e]*dodecakisbenzene*; see also Fig. 2. In this nomenclature the symbols d and e [in brackets] reflect the linear or angular annelation of a benzenoid ring, respectively.

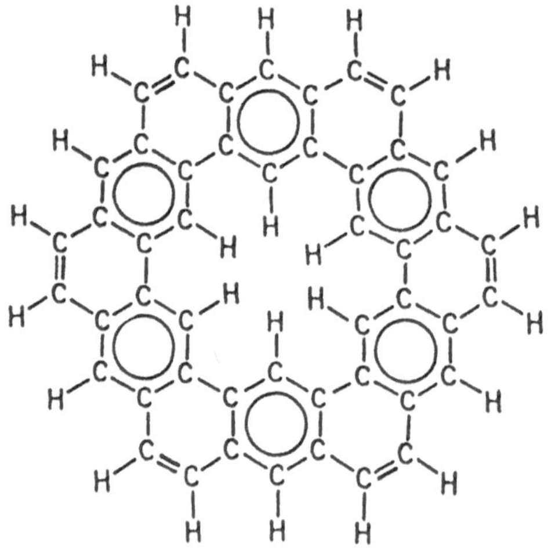

Fig. 1.2. The molecule C_1 of Fig. 1: *cyclo*[d.e.d.e.d.e.d.e.d.e.d.e]*dodecakisbenzene* or *kekulene*.

Actually the first report on the attempts to synthesize C_1 (Figs. 1 and 2) was given in 1965 by H. A. Staab at the Annual Meeting of "Gesellschaft Deutscher Chemiker" in Bonn, the Kekulé Centennial. At this occasion the molecule C_1 was named *kekulene*. Not before thirteen years later (Diederich and Staab 1978) the first successful large-scale synthesis of *kekulene* was reported.

After this historical synthesis of the "greenish-yellow microcrystals", "with its extreme insolubility in solvents of all kinds" (Diederich and Staab 1978) a number of physical properties of *kekulene* have been established, including the molecular structure and spectroscopic (electronic) properties (Diederich and Staab 1978; Krieger et al. 1979; Schweitzer et al. 1982; Staab and Diederich 1983; Staab, Diederich, Krieger and Schweitzer 1983). Theoretical studies are also available, including molecular geometry, aromaticity, magnetic susceptibility, [1]H-chemical shifts and zero-field splitting (Ege and Fischer 1967; Ege and Vogler 1972a; 1972b; Vogler 1979; 1980; 1985; Randić 1983). An early study on diamagnetic anisotropy is due to McWeeny (1951). A complete set of vibrational frequencies have been calculated (Cyvin SJ, Brunvoll, Cyvin and Brendsdal 1988).

1.2.2 *Other Cycloarenes*

An article series in five parts, entitled CYCLOARENES, A NEW CLASS OF AROMATIC COMPOUNDS, has been published by H.A. Staab with collaborators (Staab and Diederich 1983; Staab, Diederich, Krieger and Schweitzer 1983; Staab, Diederich and Čaplar 1983; Staab and Sauer 1984; Funhoff and Staab 1986).

The synthesis of the C_2 molecule of Fig. 1, viz. *cyclo*[d.e.d.e.e.d.e.d.e.e]-*decakisbenzene*, has also been attempted several times (Jenny and Peter 1965a; 1965b; Peter and Jenny 1966; Staab, Diederich and Čaplar 1983; see also the comment by Clar et al. 1981). The goal was reached eight years after the synthesis of *kekulene* (Funhoff and Staab 1986). Also for C_2 (Fig. 1) a number of theoretical results are available (see the references under *kekulene*), including calculated vibrational frequencies (Brendsdal et al. 1988).

Finally we mention that syntheses of *cyclo*[d.e.e.d.e.e.d.e.e]*nonakisbenzene* (C_3 of Fig. 1) and *cyclo*[d.e.d.e.d.e.d.e.d.e]*decakisbenzene* (C_4 of Fig. 1) are under way (Staab and Sauer 1984). It is again referred to the references under *kekulene* for theoretical works. In particular, see Cyvin SJ, Brunvoll and Cyvin (1988) for a complete set of calculated vibrational frequencies for C_3.

1.3 CORANNULENES

The *cycloarenes* belong to a wider class of conjugated hydrocarbons called *corannulenes*. They are characterized by two macrocyclic rings of carbon atoms (*annulene* systems; see below), one inner being completely embraced by an outer. These two cycles are connected by a number of (radial) bonds. This concept, which was introduced by Hellwinkel (1970), has been widely adopted by others (Ege and Vogler 1972a; 1972b; Vogler 1980; Agranat et al. 1980; Randić and Trinajstić 1984). An alternative name is "circulenes" (Dopper and Wynberg 1972; 1975).

A great variety of molecules belongs to the class of *corannulenes*. Figure 3 shows three examples including *coronene* (O_2). They all consist of a number of six-membered (benzenoid) rings around a cavity of varying size, which is reflected in the nomenclature [5]*circulene* (O_1), [6]*circulene* (O_2) and [7]*circulene* (O_3).

The name *corannulene* was originally given as a trivial name to O_1 (Fig. 3) after it had been synthesized (Barth and Lawton 1966; 1971). *Coronene*, O_2, is long known (see, e.g. Clar 1964). Also the third molecule of Fig. 3, O_3, has been synthesized (Yamamoto et al. 1983). A synthesis of [8]*circulene* has been attempted (Thulin and Wennerström 1976).

Corannulenes are not confined to six-membered rings around the cavity. Some of the references cited above contain a plethora of examples of non-benzenoids with different sizes of the rings, but we shall not consider such systems here.

We wish to conclude this section by some remarks on *pyrene* and *coronene*, which both have been reckoned among *corannulenes* (Ege and Vogler 1972a; 1972b; Vogler 1980). Figure 4 shows how two forms of the dualist (or a dualist-like figure) can distinguish between say [4]*circulene* and *pyrene* on one hand, as well as [6]*circulene* and *coronene* on the other. The representations in terms of hexagons, however, are unique; see the right-hand drawings in Fig. 4. It is also a fact that there only exists one chemical compound *pyrene* ($C_{16}H_{10}$) and one *coronene* ($C_{24}H_{12}$). It is most appropriate to consider *coronene* as a system of seven hexagons instead of one with a "cavity". With regard to *pyrene* it seems appropriate to characterize it at least as a degenerate *corannulene*.

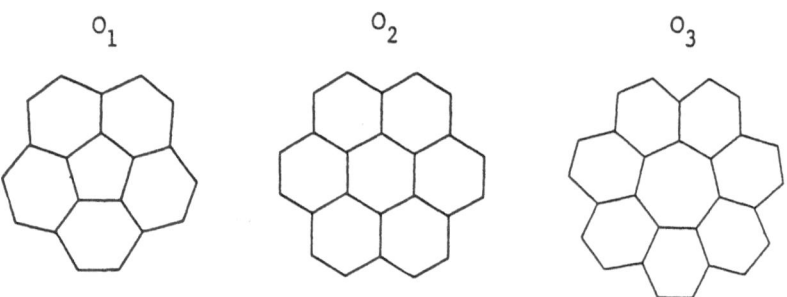

O_1 O_2 O_3

Fig. 1.3. Three *corannulenes*: $C_{20}H_{10}$ (O_1), $C_{24}H_{12}$ (O_2) and $C_{28}H_{14}$ (O_3).

1.4 OTHER MACROCYCLIC CONJUGATED HYDROCARBONS

1.4.1 *Introduction*

Very much work has been done on macrocyclic conjugated hydrocarbons in chemistry. Many of these compounds are of interest here, although they do not correspond to coronoid systems. In the following we shall describe *annulenes* with mentioning of *annulenoannulenes*, and further the wide class of annelated *annulenes*, to which one may reckon the macrocyclic *polyphenylenes*.

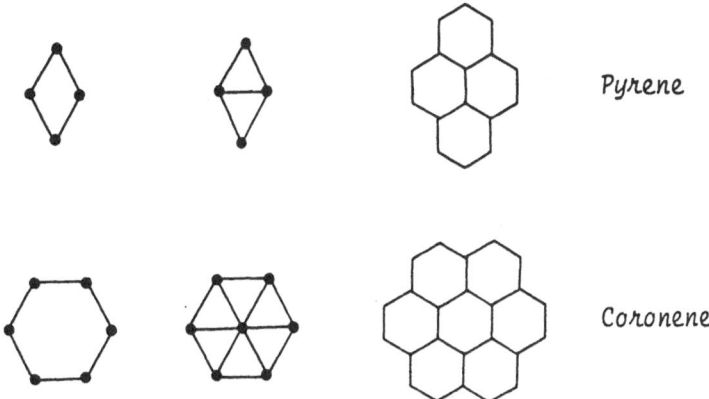

Pyrene

Coronene

Fig. 1.4. *Pyrene* and *coronene* considered as *corannulenes* or as pericondensed benzenoids.

1.4.2 *Annulenes*

The term *annulene* was coined by Sondheimer and Wolovsky (1962). These compounds are macrocyclic *polyenes*, thus containing a cycle of conjugated carbon-carbon bonds. The concept allows for both even- and odd-numbered cycles. A great number of reviews or lecture reports on *annulenes* in their diversity of forms are available; here we give reference to only some of them: Sondheimer (1963; 1971); Haddon et al. (1971); Sondheimer (1972; 1974); Müllen (1984).

Here we shall only consider the type of *annulenes* where the number of carbon-carbon bonds (equal to the number of carbon atoms) is even. The distinction between [4k + 2]- and [4k]-cycles is both chemically and theoretically important. The bracketed number, where k is an integer, indicates the number of bonds (or atoms). In short, the [4k + 2]- and [4k]-cycles result in aromatic and anti-aromatic compounds, respectively.

Benzene is a trivial *annulene* and may as such be designated [6]*annulene*. It is of the [4k + 2] type with k=1. Also [8]*annulene* or *cyclooctatetraene* is long known (see, e.g. the reviews: Sondheimer 1963; 1974). This is a nonplanar molecule and belongs to the [4k] type.

Figure 5 shows four *annulene* systems, where both types of cycles are represented, viz. [4k + 2] in A_1, A_2 and A_4, but [4k] in A_3. The corresponding molecules are known. For [14]*annulene* (A_1), see Sondheimer and Gaoni (1960); Gaoni and Sondheimer (1964). Also the larger *annulenes*, viz. [18]*annulene* (A_2), [24]*annulene* (A_3) and [30]*annulene* (A_4) have been prepared (Sondheimer, Wolovsky and Amiel 1962). An earlier report on the synthesis of A_2 exists (Sondheimer and Wolovsky 1959).

A theoretical work on aromatic stabilization of *annulenes* by Gutman, Milun and

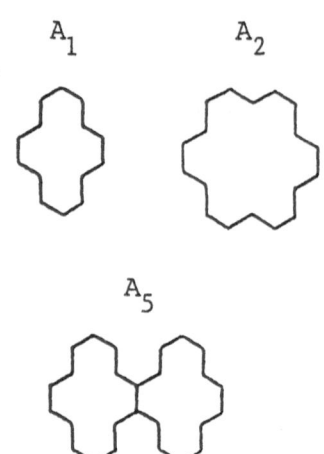 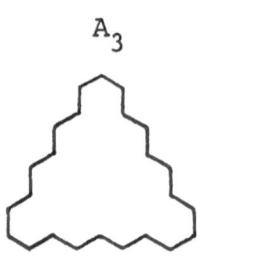 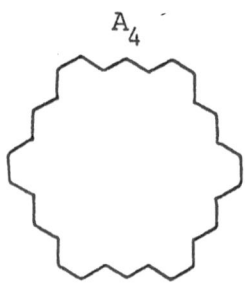

Fig. 1.5. [14]Annulene (A₁), [18]annulene (A₂), [24]annulene (A₃), [30]annulene (A₄) and [14]annuleno[14]annulene (A₅).

Trinajstić (1971) is worth mentioning. Vogler's (1980) calculations of zero-field splitting parameters include some *annulenes*. Stollenwerk et al. (1983) have calculated magnetic susceptibilities and resonance energies for some molecules including [14]*annulene* and [18]*annulene*; for the former *annulene* the magnetic susceptibility was also determined experimentally.

1.4.3 *Annulenoannulenes*

An *annulenoannulene* contains two (macro)cyclic *polyenes*. The long known *naphthalene* (see, e.g. Clar 1964) is a trivial example. Figure 5 contains a non-trivial example, viz. [14]*annuleno*[14]*annulene* (A₅), of which some derivatives have been synthesized (Cresp and Sondheimer 1975; 1977). The latter paper (Cresp and Sondheimer 1977) also deals with larger *annulenoannulene* derivatives.

Here we mention a theoretical work on [1]H-chemical shifts of some *annuleno-annulenes* (Vogler and Ege 1976) and one on the aromaticity (Hess et al. 1978).

1.4.4 *Annelated* Annulenes

Different types of additions to *annulenes* give compounds which have been referred to as annelated *annulenes*.

In a special type of the compounds in question one or more catacondensed benzenoids are fused to the *annulene* so that they share exactly one edge each with the *annulene*. Figure 6 shows twenty-two examples. *Benzo*[d][14]*annulene* (Q₂) was obtained (Meissner et al. 1976; Staab, Meissner, Weinacht and Gensler 1979) while the investigators tried to synthesize *benzo*[a][14]*annulene* (Q₁). Both of the isomers (Q₁ and Q₂) were included in the extensive work of Stollenwerk et al. (1983) with calculated magnetic susceptibilities and resonance energies. For the synthesized molecule (Q₂) these investigators also reported an experimental value of the

magnetic susceptibility. Furthermore, this molecule (Q_2) was considered by Randić
(1983) in his discussion of the role of Kekulé structures in aromaticity and anti-
aromaticity. Only theoretical studies are available for *dibenzo[14]annulenes*.
Stollenwerk et al. (1983) considered all the five isomers of them, viz. $Q_3 - Q_7$.
Three of these molecules, viz. Q_3, Q_5 and Q_6, were also considered by Randić (1983),

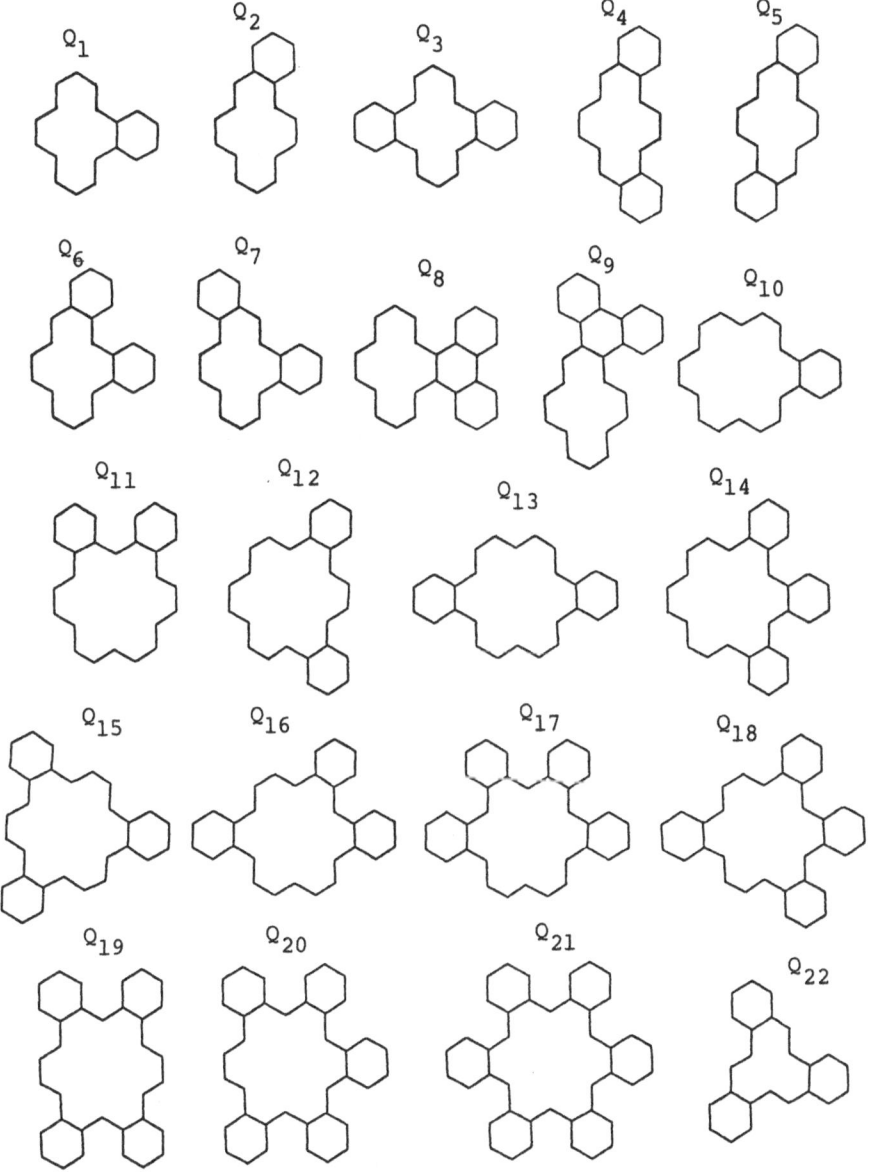

Fig. 1.6. Nine annelated [14]*annulenes* (Q_1-Q_9), twelve annelated [18]*annulenes*
(Q_{10}-Q_{21}) and one annelated [12]*annulene* (Q_{22}).

while in a later work, Vogler and Trinajstić (1988) applied the theory of conjugated circuits to *dibenzo*[d,k][14]*annulene* (Q_5). The *phenanthro*[14]*annulene* identified as Q_8 in Fig. 6 has been synthesized (Staab, Meissner, Weinacht and Gensler
1979), while its isomer Q_9 is more speculative. The synthesized form (Q_8) is treated in the work of Stollenwerk et al. (1983) with the experimental magnetic susceptibility included. These authors have also treated all the possible *(poly)benzo*-
[18]*annulenes* (Q_{10} - Q_{21}). *Tribenzo*[a,d,g][18]*annulene* (Q_{14}) was also treated by
Vogler and Trinajstić (1988). Calculated ^{1}H-chemical shifts for *benzo*[a][18]*annulene* (Q_{10}) were reported by Ege and Vogler (1975), prior to the synthesis of this
molecule (Meissner, Gensler and Staab 1977; Staab, Meissner and Gensler 1979).
Finally in this category we mention *tribenzo*[a,e,i][12]*annulene* (Q_{22}), which also
has been synthesized (Staab, Graf and Junge 1966; Staab, Graf, Doerner and Nissen
1971).

In an other type of annelated *annulenes*, the macrocyclic *polyphenylenes*, ben-
zene units are bonded into a macrocycle. Both ortho- meta, and para bondings may
occur. Figure 7 shows as examples the chemically known molecules (P_1) *hexa-m-phenylene* (Staab and Binnig 1964; 1967a), (P_2) *octa-m-phenylene* (Staab and Binnig 1967a)
and an ortho- and para bonded *nonaphenylene* (Meyer and Staab 1969). Also *penta-m-
phenylene* (Staab and Binnig 1967b) and *deca-m-phenylene* (Staab and Binnig 1967a;
1967b) have been prepared. The bond lengths in *penta-m-phenylene*, P_1 and P_2, along
with *hepta-m-phenylene*, have been studied theoretically by Ege and Fischer (1967).
Mass spectral studies of *penta-m-phenylene*, P_1, P_2 and *deca-m-phenylene* have been
reported (Staab and Wünsche 1968), and somewhat later the crystal- and molecular
structure investigations of *penta-m-phenylene* an P_1 (Irngartinger et al. 1970).

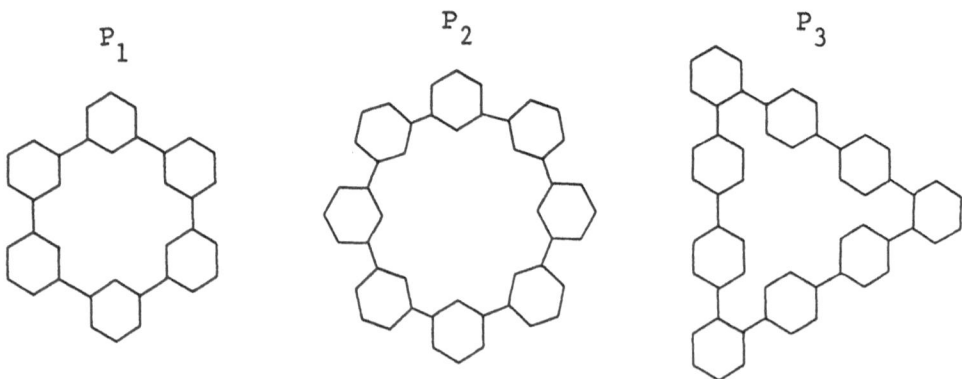

Fig. 1.7. Hexa-m-phenylene (P_1), *octa-m-phenylene* (P_2), and an ortho-para bonded
nonaphenylene (P_3).

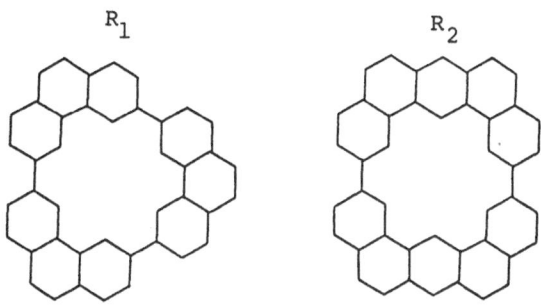

Fig. 1.8. Two annelated *annulenes* related to the macrocyclic *polyphenylenes*. R_1 is *triphenanthro*[abcde,ghijk,mnopq][18]*annulene*.

Triphenanthro[abcde,ghijk,mnopq][18]*annulene* (R_1 of Fig. 8) was prepared by Staab et al. (Staab and Bräunling 1965; Staab, Bräunling and Schneider 1968) and subjected to mass spectral studies (Staab and Wünsche 1968). This molecule is included in the theoretical studies of Ege and Fischer (1967) and of Stollenwerk et al. (1983) . The latter work includes the experimental magnetic susceptibility of R_1. This molecule is related to the macrocyclic *polyphenylenes*, as also is the case with the other molecule (R_2) depicted in Fig. 8. Vogler (1979) calculated the [1]H-chemical shifts of the latter molecule (R_2).

The macrocyclic *polyphenylenes* as well as R_1 and R_2 may be described as *benzene* units or larger benzenoids bonded together into a macrocycle by isolated bonds. Similar annelated *annulenes* exist where the linkages consist of chains of bonds. A collection of such compounds is shown in Fig. 9 (where in five examples, viz. S_3, S_8, S_{12}, S_{14} and S_{16}, only one benzenoid in each is distinguished). The molecule S_6 (see Fig. 9) was prepared relatively early by Cram and Dewhirst (1959). It is mentioned in a work by DuVernet et al. (1978) and, together with S_1, S_2 and S_4, by Otsubo et al. (1978). The preparations of *dibenzo*[ab,de][18]*annulene* (S_7) and *phenanthro*-[cdefg][18]*annulene* (S_8) have been reported (Meissner et al. 1973; Staab, Meissner and Meissner 1976), and both molecules are included in the theoretical work of Ege and Vogler (1975). The latter molecule (S_8) is found with experimental magnetic susceptibility in the work of Stollenwerk et al. (1983), who also considered theoretically S_3, S_5, S_9 and S_{10}. Vögtle and Staab (1968) reported the synthesis of S_{11}. This molecule, together with S_{12}, plays a role in the synthesis of *kekulene* (Staab and Diederich 1983). The compounds S_{13} and S_{14} were prepared during the studies towards a synthesis of *cyclo*[d.e.d.e.e.d.e.d.e.e]*decakisbenzene* (Staab, Diederich and Čaplar 1983). Similarly, S_{15} and S_{16} have been prepared during an attempted synthesis of *cyclo*[d.e.e.d.e.e.d.e.e]*nonakisbenzene*.

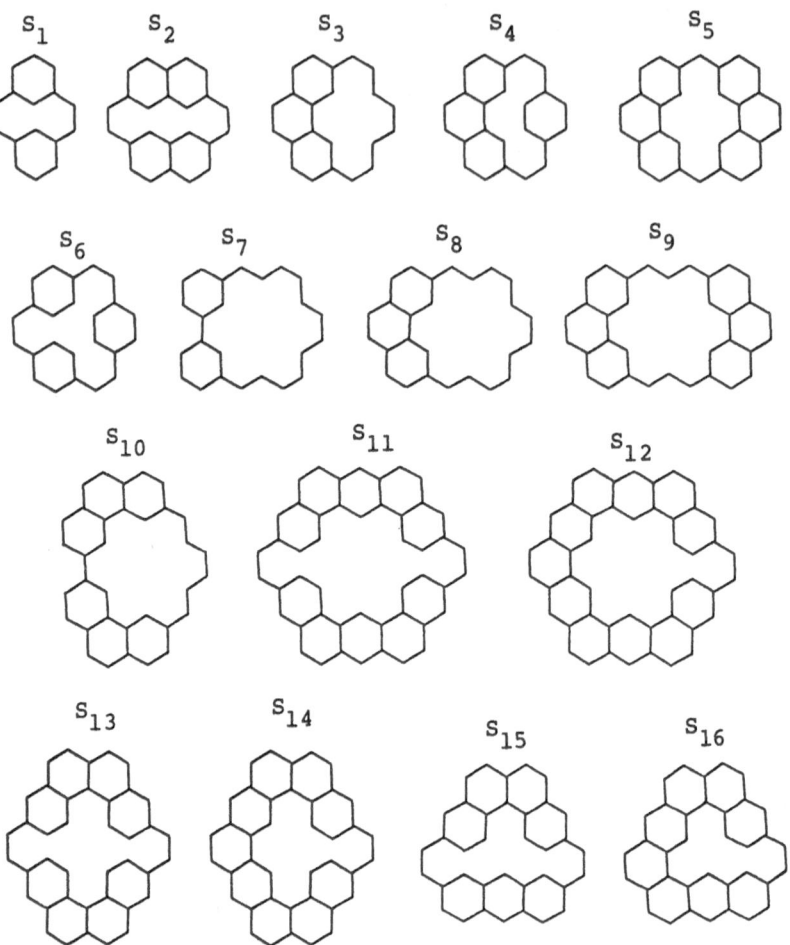

Fig. 1.9. Sixteen annelated *annulenes*.

Herndon (1974) studied the bond orders and bond lengths in *annulenes* and some annelated *annulenes* including *penta-m-phenylene*, *hexa-m-phenylene* (P_1), R_1 (cf. Fig. 8) and S_8 (Fig. 9). Randić (1975), in his derivation of Pauling bond orders, analysed the Kekulé structures of S_3 (Fig. 9). *Hexa-m-phenylene* (P_1) is included in the studies of characteristic polynomials and eigenvalues by Dias (1988).

Chapter 2

CORONOIDS AND CORONOID-LIKE SYSTEMS

2.1 CORONOID SYSTEM

2.1.1 *Basic Definitions*

A coronoid system (or shortly *coronoid*) is a geometric figure which, like a benzenoid, consists of congruent regular hexagons in a plane. Also, like the hexagon, the concepts of vertex, edge and isomorphic/nonisomorphic systems are straightforwardly transferred from the theory of benzenoids. The reader is referred to a recent monograph entitled INTRODUCTION TO THE THEORY OF BENZENOID HYDROCARBONS, which contains a chapter on coronoids (Gutman and Cyvin 1989b).

Loosely speaking a coronoid is a "benzenoid with a hole".

The strict definition of a benzenoid in terms of a cycle on a hexagonal lattice is easily adapted to a coronoid. Assume two cycles, C' and C", where C" is completely embraced by C'. The vertices and edges on C" and its interior should consist of at least two hexagons. This part of the hexagonal lattice is referred to as the *corona hole*. A coronoid consists of the vertices and edges on C' and C" as well as in the interior of C', but outside C". An illustration is given below.

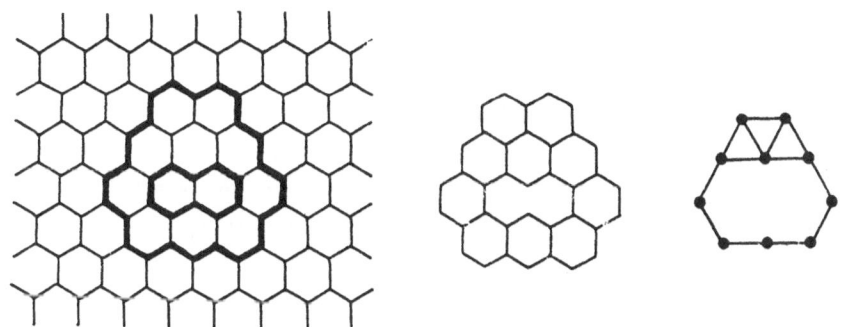

The representation of a coronoid in terms of a dualist (previously called "dualist graph") may be very convenient and is exemplified above. In a dualist each vertex represents a hexagon.

The cycles C' and C" are the *outer-* and *inner perimeter* of the coronoid, respectively.

The above definition applies, strictly speaking, only to *single coronoids*, i.e. those with one and only one hole. Throughout this book only single coronoids are considered, and it shall be tacitly assumed that the term "coronoid" actually refers to "single coronoid" if nothing else is stated.

In order to prepare for an alternative definition of coronoids we shall define the class of primitive coronoids.

A *primitive coronoid* consists of a single chain of hexagons in a circular (macrocyclic) arrangement. Thus any vertex of degree three on one perimeter (say C') is adjacent to a vertex of degree three on the other perimeter (C"). Edges not belonging to the perimeter of a primitive coronoid are called *radial edges*.

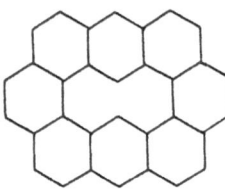

Fig. 2.1. The smallest coronoid system.

The smallest coronoid is a unique system of eight hexagons (h=8). It is a primitive coronoid (see Fig. 1).

Now it is inferred that all coronoids with h+1 hexagons are constructed by (a) adding one hexagon to all coronoids with h hexagons and (b) including the primitive coronoids with h+1 hexagons.

The addition of hexagons is taken in the same sense as in the corresponding construction of benzenoids. For the sake of convenience we reproduce also here the five addition modes; see Fig. 2.

It is emphasized that allowance is made for an addition of a hexagon to the inner perimeter (as well as the outer), if such an addition is possible.

The above characterization is not an effective definition unless we have a way to construct all primitive coronoids for a given number of hexagons. This may be achieved in different ways. One way is to allow for an addition of a hexagon to a single (catacondensed) unbranched chain which closes the chain. This addition has been referred to as *corona-condensation* or "cyclic cata-condensation" (Polansky and Rouvray 1977). Another way would be to take benzenoids with increasing numbers of hexagons as corona holes in a systematic way. We shall not go into details of this procedure here since it is to be treated in a subsequent section as a practical method for generation and enumeration of primitive coronoids.

2.1.2 *Terminology*

The term "coronoid" was introduced by Brunvoll, Cyvin BN and Cyvin (1987a). More or less synonymous designations which have been transferred from the hydrocarbon chemistry (see also Sections 1.1 and 1.3), are "coronaphene" (Jenny and Peter 1965a; Balaban and Harary 1968), "(true) circulene" (Dopper and Wynberg 1972; 1975; Dias 1982; Knop, Szymanski, Jeričević and Trinajstić 1984) and "coronafusene" (Balaban 1982).

Balaban and Harary (1968) also suggested in a footnote the term "corona-con-

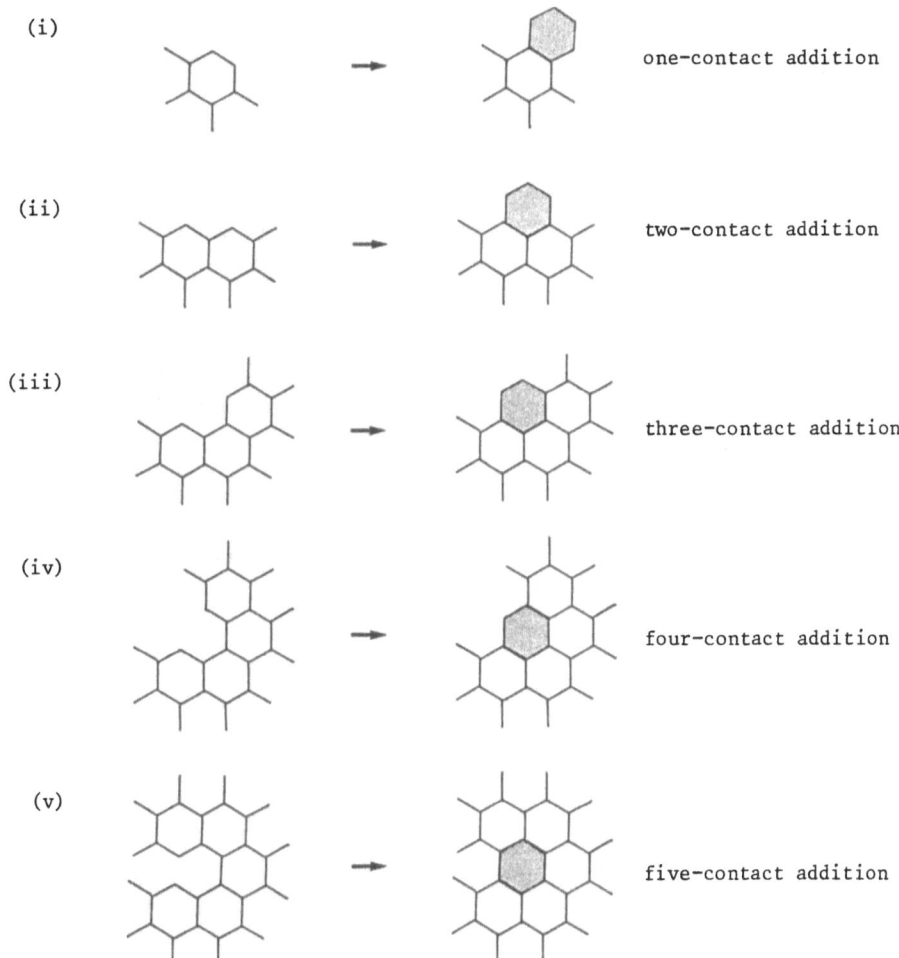

Fig. 2.2. Five types of addition of a hexagon. The added hexagon is grey. The pen-
dent lines symbolize hexagons which may, but need not exist.

densed" system, which should fit into the following definitions based on dualists:
a system is catacondensed when its dualist is a tree (no rings), pericondensed when
it contains three-membered rings, i.e. triangles, and corona-condensed when it contains a larger than three-membered ring. The term corona-condensed system has been
adopted by others (e.g. Polansky and Rouvray 1977). The main purpose of introducing
the term coronoid (Brunvoll, Cyvin BN and Cyvin 1987a) was to make it possible to
speak about catacondensed and pericondensed coronoids (for definitions, see a subsequent section) in analogy with catacondensed and pericondensed benzenoids.

In the remainder of the present book we shall avoid the terms "circulene",
"corannulene" (see Section 1.3), "corona-condensed" system, "coronafusene", "coronaphene" and "cycloarene" (see Sections 1.1 and 1.2).

2.2 INTERLUDE

2.2.1 *Coronoid Systems and Coronoid Hydrocarbons*

There is an obvious correspondence between a coronoid system and a coronoid hydrocarbon (chemically known or not). It is sufficient to point at the system of hexagons (C_1) in Fig. 1.1 and the formula of *kekulene* in Fig. 1.2.

A vertex of degree two (resp. three) of a coronoid system corresponds to a secondary (resp. tertiary) carbon atom of the appropriate coronoid hydrocarbon.

In order not to depart too far from chemistry we shall also call particular coronoid systems by the corresponding chemical names. Thus the primitive coronoid C_1 will be referred to as "*kekulene*". In the same way the benzenoids of Fig. 1.4 are called "*pyrene*" and "*coronene*".

2.2.2 *Kekulé Structures*

The so-called *1*-factor of a benzenoid or coronoid system corresponds to the notion of Kekulé structure (see, e.g. Cyvin and Gutman 1988) from organic and physical chemistry. Here we shall use the term "Kekulé structure" (as in the case of benzenoids) also about the coronoid systems, and when appropriate indicate a Kekulé structure by single and double bonds; see Fig. 3.

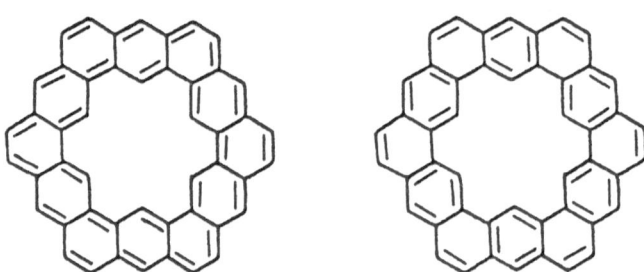

Fig. 2.3. Two Kekulé structures (out of 200) for *kekulene*.

The number of Kekulé structures for a given coronoid (or benzenoid) is usually designated by the symbol K. It is also called the Kekulé structure count.

2.3 NON-CORONOID SYSTEMS

It was pointed out above that a coronoid system corresponds (in an obvious way) to a coronoid hydrocarbon. On the other hand there are chemical formulas which it is natural to assign to coronoid hydrocarbons, but do not correspond to coronoid systems according to the definition in Paragraph 2.1.1. A typical example is furnished by C_4 of Fig. 1.1 (cf. also Section 1.1). We shall refer to C_4 as a non-coro-

noid system (or *non-coronoid*) without any need for a precise definition of the term.
All the examples of Figs. 1.3 - 1.9 may be associated (more or less naturally) to
non-coronoid systems.

A non-coronoid system falls under the category of non-benzenoid systems (but
neither this term seems to have been precisely defined anywhere).

The notion of Kekulé structures is applicable to non-coronoid/non-benzenoid as
well as coronoid/benzenoid systems.

Among the non-coronoid systems there are systems closely related to coronoids
which are often encountered in the studies of real coronoids. They should be defined
more strictly and fall under two categories: degenerate coronoid systems (or *degenerate coronoids*) and quasi-coronoid systems (or *quasi-coronoids*). These types are
treated in the next two sections.

A degenerate coronoid or quasi-coronoid may (like a coronoid) be superimposed
on a hexagonal lattice (cf. Paragraph 2.1.1).

2.4 DEGENERATE CORONOID SYSTEMS

A degenerate coronoid is a connected part of a coronoid containing edge(s) not
belonging to hexagons.

It may, for instance, be a system with exocyclic bonds, either attached to a
coronoid or connecting a coronoid with one or more benzenoids. The similar types are
well known from the studies of benzenoids (Trinajstić 1983; Cyvin and Gutman 1988).
Figure 4 shows some illustrative examples (the three first drawings). These systems
can appropriately be termed *annelated polyenes*.

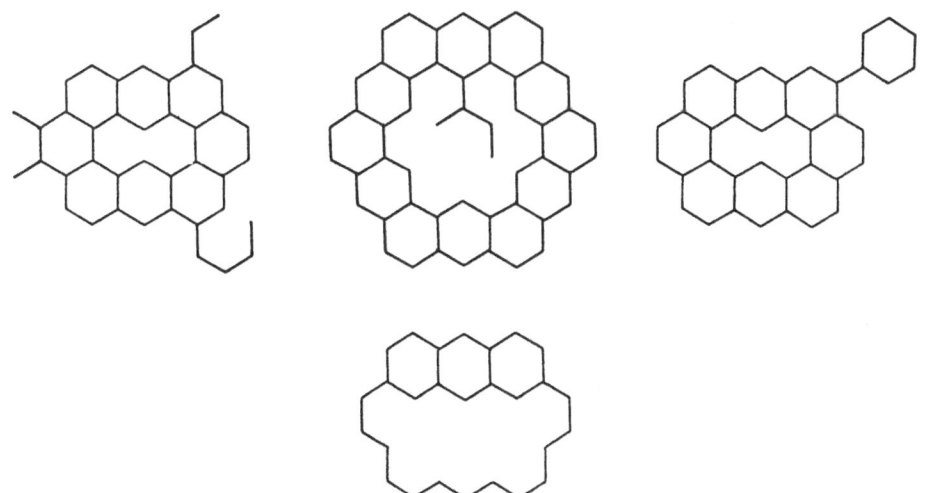

Fig. 2.4. Degenerate coronoids: annelated *polyenes* (top row) and an annelated
annulene (bottom drawing).

It may also happen that the edges not belonging to hexagons are a part of a macrocycle (see the bottom drawing of Fig. 4). This is a case which may be interpreted as a coronoid-like system where parts of the outer and inner perimeter coalesce. Additional examples are furnished by the systems corresponding to the hydrocarbons P_1 and P_3 (not P_2) of Fig. 1.7, and all the systems of Figs. 1.6, 1.8 and 1.9. We shall use the term annelated *annulenes* (cf. Paragraph 1.4.4) also about this kind of coronoid systems (and not only coronoid hydrocarbons).

Annulenes and *annulenoannulenes* among degenerate coronoids (cf. Fig. 1.5) represent the extreme cases of systems which do not possess any hexagon.

Of course there exist hydrocarbons among *annulenes* which can not be superimposed on a hexagonal lattice and therefore do not correspond to a degenerate coronoid. All odd-membered cycles and [8]*annulene* (cf. Paragraph 1.4.2) are examples.

Special attention to the *annulenes* among degenerate coronoids was paid by Balaban (1971), who studied their construction and enumeration. These systems consist simply of a closed cycle on the hexagonal lattice, where all edges meet at angles of 120°. Both [4k + 2]- and [4k]-cycles are allowed for. The mentioned author, however, considered also cases with overlapping edges, which we do not reckon among degenerate coronoids. Overlapping edges are encountered among helicenic systems described under the next section.

2.5 QUASI-CORONOID SYSTEMS

2.5.1 *Coronene and Some of Its Derivatives*

It is agreed that the corona hole of a coronoid must have a size of at least two hexagons. Nevertheless in Fig. 1.4 it was demonstrated how *coronene* can be interpreted as a coronoid-like system with a hole of one hexagon only. As such this system of six hexagons will presently be referred to as a quasi-coronoid. It plays a quite important role in some parts of the studies of real coronoids.

The concept of quasi-coronoids may be extended to any derivative of *coronene*, produced by arbitrary additions of hexagons. It would lead us far astray to consider all such systems, as for instance *ovalene* (Fig. 5). Other cases may be of a greater interest, e.g. *circumcoronene* (see Fig. 5).

2.5.2 *Helicenic Systems*

Helicenic systems are well known from the studies of benzenoids. They are characterized by the presence of overlapping edges. We shall refer to such systems as helicenic quasi-coronoids if something like a corona hole can be identified. Figure 6 (top) shows an example where the helicenic character is caused by an appendage to a coronoid. In the other example (Fig. 6, bottom) a deformed corona hole is essential for creating the helicenic system.

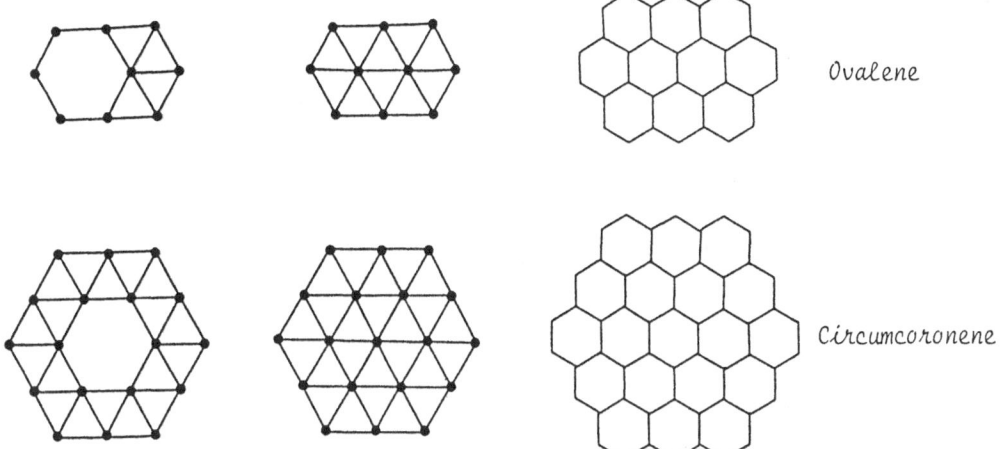

Fig. 2.5. *Ovalene* and *circumcoronene* considered as quasi-coronoids and as peri-condensed benzenoids.

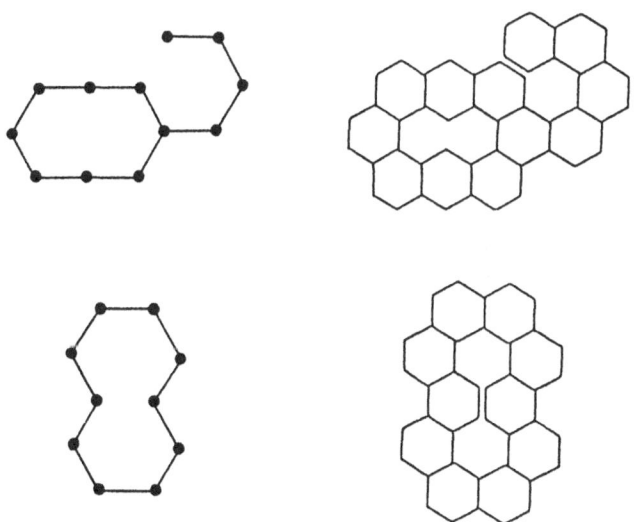

Fig. 2.6. Two examples of helicenic quasi-coronoids: dualists and hexagon representations.

2.6 CONCLUDING REMARKS

It seems that the coronoids can no longer be circumvented in relevant reviews and monographs, although benzenoids may be the main topic (Gutman 1982; Balaban 1982; 1985; 1988; 1989; Randić 1983; Trinajstić 1983; Trinajstić et al. 1986; Cyvin and Gutman 1988; Gutman and Cyvin 1989b). It is true, however, that coronoids sometimes are mentioned only in order to exclude them. As Gutman (1982) says about different types of non-benzenoids including a coronoid: "[they] will not be considered here". Also, in a study of simply connected benzenoid graphs Polansky and Gutman (1979) make it clear explicitly that coronoids are excluded. Likewise in Cyvin SJ, Brunvoll and Cyvin (1989b). On the other hand, it seems to be an exaggeration when Trinajstić et al. (1986) say: "Many such structures [having "holes"] are experimentally known, the prototype being Kekulčne".

Many works deal with theories developed especially for coronoids (cf. references in Chapter 1), or with more general theories where a coronoid is used as example. Davidson (1981) included *kekulene* as one of many examples (the other not coronoids) in his spectral analysis of graphs. *Kekulene* in particular has frequently been used as example: for instance in studies of aromatic sextets (Aihara 1976; Ohkami et al. 1981; Gutman and El-Basil 1984; Zhang et al. 1990), in calculation of total π-electron energy (Cioslowski 1985), and in the computer-generation of a topological polynomial called matching polynomial (Ramaraj and Balasubramanian 1985; Balasubramanian 1987). Randić (1986), in a statistical approach to resonance energies, generated a random sample out of the 200 Kekulé structures for *kekulene*. Dias (1988) included *kekulene* in his studies of characteristic polynomials and eigenvalues.

Also other coronoids than *kekulene* have been used as examples in computational chemistry: a program for generation and enumeration of Kekulé structures was exemplified by *cyclo*[d.e.d.e.e.d.e.d.e.e]*decakisbenzene* (Džonova-Jerman-Blažič and Trinajstić 1982), while Tošič and Kovačević (1988) only mention the existence of coronoids by passing, but provide an illustration as example. *Circumkekulene* along with another related coronoid were used to exemplify a computation of reduced cycle indices (Balasubramanian 1989). He WC and He (1987) used a coronoid (other than *kekulene*) as example in their study of aromatic sextets; in this connection also another work by He WJ and He (1986) and one by Ohkami (1990) should be mentioned. Ohkami et al. (1981) analysed the aromatic sextets in *cyclo*[d.e.e.e.d.e.e.e]*octakisbenzene*, while the same system was considered by Randić et al. (1989) in connection with a study of helicenic quasi-benzenoids. Furthermore, Randić et al. (1988) reported a Kekulé structure count for a helicenic quasi-coronoid. The peak-valley path method (He WJ and He 1987; He WC and He 1990c) is claimed to be valid for coronoids as well as for benzenoids.

PART II

ANATOMY

Chapter 3

ANATOMY: GENERAL TREATMENT

3.1 INTRODUCTION

In this chapter some fundamental characteristics of the coronoid systems are described. Special attention is paid to several invariants and relations between them. Furthermore, several classification schemes for coronoids are put forward.

The monograph quoted in the preceding chapter (Gutman and Cyvin 1989b) should be consulted as a general reference, especially the chapter with the same name (ANATOMY) therein. A pioneering work on some topological properties of coronoid systems is due to Polansky and Rouvray (1976a); (1977). A more recent, significant paper with relevance to this topic is due to Hall (1988). More rudimentary results are found in Dias (1982) and in He WJ and He (1986); cf. also Cyvin SJ and Brunvoll (1989). Several papers have appeared on enumeration and classification of coronoids. They are cited in subsequent chapters. Here we only mention an original work in which some details of the classification of coronoids were exposed (Cyvin SJ, Cyvin and Brunvoll 1987).

3.2 BASIC CONCEPTS

3.2.1 *Hexagon Modes and Some Structural Features*

A hexagon of a coronoid system may be assigned to one of twelve modes, exactly as in the case of benzenoids. For the sake of convenience we reproduce also here (Fig. 1) the picture which defines these modes and indicates the adopted notation. An example is provided in Fig. 1.

Five of these modes are addition modes as described in Paragraph 2.1.1. We are now in the position to identify the modes of the added hexagon (cf. Fig. 2.2) in terms of the notation of Fig. 1. They are: (i) L_1, (ii) P_2, (iii) L_3, (iv) P_4 and (v) L_5.

In Paragraph 2.1.1 also the corona-condensation is defined. Under this process the added hexagon may acquire the mode L_2 or A_2.

A free edge is a (2,2)-edge, thus indicating that it implies two vertices of degree two. Similarly we may indicate a fissure, bay, cove and fjord by (2,3,2), (2,3,3,2), (2,3,3,3,2) and (2,3,3,3,3,2), respectively. In the case of coronoids these structural features may occur either on the outer or the inner perimeter. Accordingly we may speak about *outer free edge, -fissure, -bay, -cove* and *-fjord* on one hand, and *inner free edge, -fissure, -bay, -cove* and *-fjord* on the other.

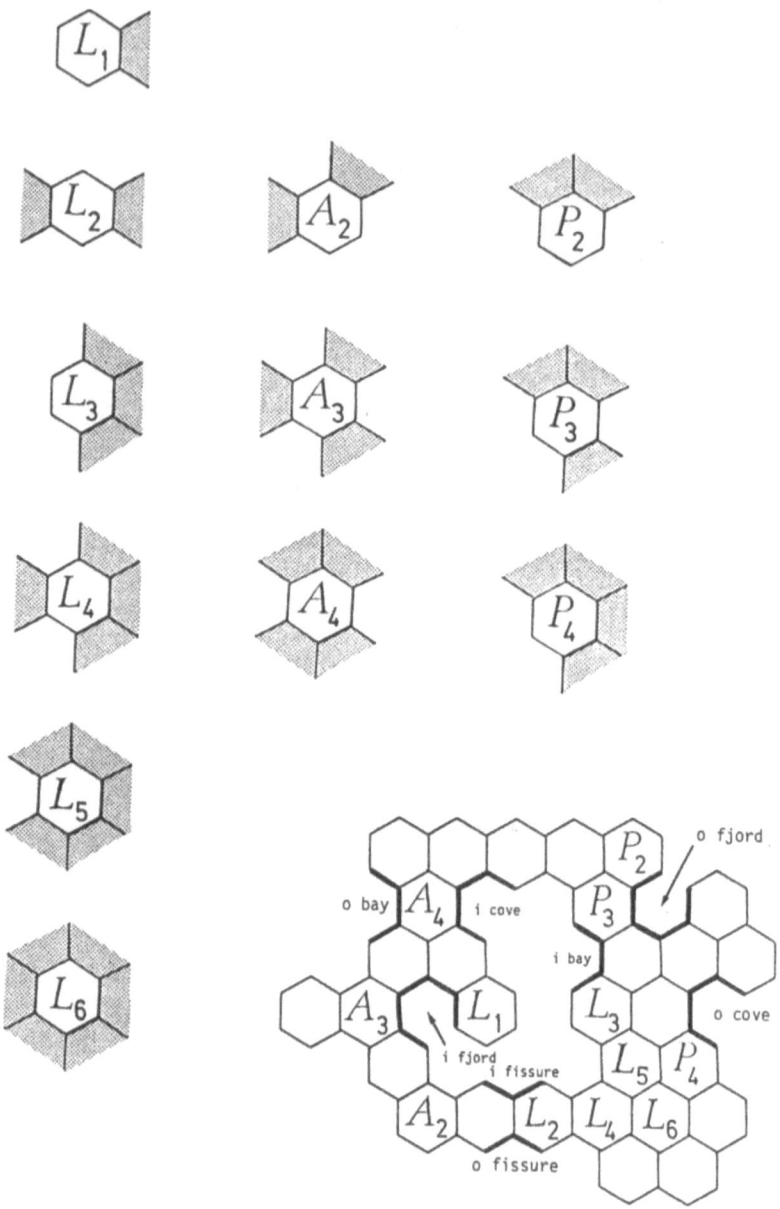

Fig. 3.1. The twelve modes of hexagons in a coronoid or benzenoid system. Definitions of some structural features of the perimeter. Abbreviations: i inner; o outer.

Figure 1 provides examples. The depicted coronoid contains also a number of free edges (fifteen outer and four inner).

After this we can also characterize the five types of addition (cf. Fig. 2.2) in the following way: (i) addition to a free edge (annelation), (ii) condensation (peri-condensation) into a fissure, and filling a (iii) bay, (iv) cove or (v) fjord.

The number of hexagons in a coronoid, like in a benzenoid, is denoted by h. Let the number of hexagons of the corona hole (interpreted as a benzenoid) be identified by the symbol h^o.

Any benzenoid with $h^o \geq 2$ may serve as a corona hole.

Convention: Throughout this book we adhere to the convention (which also frequently is used for benzenoids) that all coronoid systems are drawn so that two edges of every hexagon are vertical.

3.2.2 *Vertices and Edges. Some Invariants*

A vertex in a coronoid which is shared by three hexagons is called an internal vertex. The same definition applies to an internal vertex of a benzenoid. On the other hand, a vertex in a coronoid which is not internal, shall not be called "external" because it may lie on either the outer or inner perimeter. Instead we shall refer to it as a *boundary vertex*. A boundary vertex on the outer (resp. inner) perimeter is called an *outer* (resp. *inner*) boundary vertex. It may happen that a coronoid system has only boundary (and no internal) vertices.

Let the number of internal vertices in a coronoid, like in a benzenoid, be denoted by n_i. The quantities h and n_i are two of the so-called invariants. It is known that for a benzenoid system a number of other invariants may be expressed linearly by these two (independent) invariants.

Also for a coronoid system there are some invariants which may be expressed by h and n_i. The following relationships exist.

$$n + n_i = 4h \tag{3.1}$$

$$m + n_i = 5h \tag{3.2}$$

$$n_2 + n_i = 2h \tag{3.3}$$

$$n_3 = 2h \tag{3.4}$$

Here n is the total number of vertices, m is the number of edges, while n_2 and n_3 denote the numbers of the vertices of degree two and three, respectively. Then clearly $n = n_2 + n_3$.

The number of boundary vertices, say n_b, is $n_b = n - n_i$; hence

$$n_b + 2n_i = 4h \tag{3.5}$$

The number n_b is also the number of edges on the two perimeters. Thus it indicates the combined length of the perimeters.

We wish to go into further details and specify the numbers of outer and inner boundary vertices as additional invariants for a coronoid system. Then we have to include two more independent invariants. We have chosen h^o and n_i^o for this purpose, viz. the number of hexagons and the number of internal vertices of the corona hole, respectively; cf. Fig. 2. Let the number of outer and inner boundary vertices be identified by the symbols n_b' and n_b'', respectively. Then $n_b' + n_b'' = n_b$, and

$$n_b' + 2(n_i - n_i^o) = 4(h - h^o) - 2 \tag{3.6}$$

$$n_b'' + 2n_i^o = 4h^o + 2 \tag{3.7}$$

It is known that a benzenoid with a $[4k + 2]$-type perimeter has an even number of internal vertices (or zero), while a $[4k]$-type perimeter gives an odd number of vertices (Cvetković et al. 1974; Gutman and Cyvin 1989a). In the case of a coronoid the length of the outer and inner parameter (viz. C' and C'') is n_b' and n_b'', respectively. We can distinguish four cases:

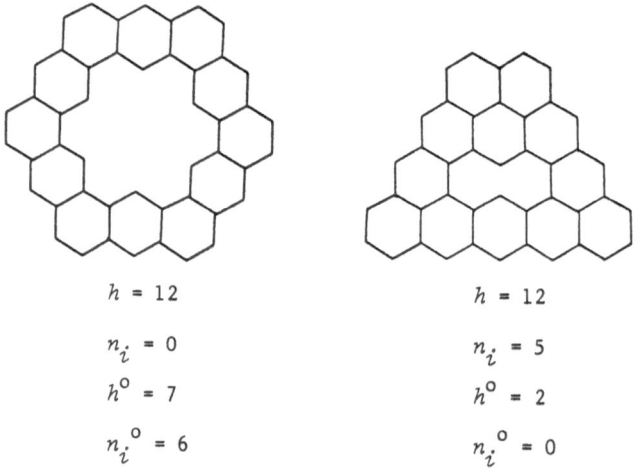

$h = 12$ $h = 12$

$n_i = 0$ $n_i = 5$

$h^o = 7$ $h^o = 2$

$n_i^o = 6$ $n_i^o = 0$

Fig. 3.2. Examples of two coronoid systems and their independent invariants.

(a) both C' and C" are [4k + 2]-cycles,

(b) both C' and C" are [4k]-cycles,

(c) C' is a [4k + 2]-cycle, but C" is [4k], and finally

(d) C' is a [4k]-cycle, but C" is [4k + 2].

The coronoid system may have n_i = 0, 2, 4, 6, ... in the cases (a) and (b), while n_i = 1, 3, 5, ... in the cases (c) and (d). The systems in Fig. 2 exemplify (a) and (d) in the left- and right-hand drawing, respectively.

All vertices of degree two are boundary vertices. They are called *outer* (resp. *inner) vertices of degree two* when they lie on the outer (resp. inner) perimeter. Let the numbers of them be designated n_2' (resp. n_2''). Then $n_2' + n_2'' = n_2$, and

$$n_2' + n_i - n_i^{\,o} = 2(h - h^o) + 2 \qquad (3.8)$$

$$n_2'' + n_i^{\,o} = 2h^o - 2 \qquad (3.9)$$

Hall (1988) pointed out the quite different properties of the outer and inner perimeter of a coronoid and gave some interesting relations for the numbers of vertices of degree two and degree three on the perimeters (see also Sachs 1984). When adapted to our notation we write

$$2n_2' - n_b' = 6 \qquad (3.10)$$

$$n_b'' - 2n_2'' = 6 \qquad (3.11)$$

Peaks and valleys in a coronoid system are special vertices of degree two and defined in the same way as in the case of benzenoids (a peak above its both first neighbours; a valley below its both first neighbours). It should be noted that peaks and valleys always occur on the outer perimeter and may (or may not) occur on the inner perimeter.

The numbers of peaks and valleys are denoted by $n_\wedge$ and n_v, respectively.

3.2.3 *Extremal Values of Some Invariants*

The upper and lower bounds of certain invariants of benzenoids were analyzed in a fine work by Harary and Harborth (1976). Some simple deductions from his formulas were made by Gutman (1982), Balaban, Brunvoll et al. (1987) and by Gutman and Cyvin (1989b).

A coronoid may be constructed for any integer $h \geq 8$. For an arbitrary h the minimum of n_i is zero, pertaining to the catacondensed systems. Therefore also the maxima of n, m and n_b, when h is given, are readily obtained as $4h$, $5h$ and $4h$, respectively. One just has to put n_i = 0 into eqns. (1), (2) and (5), respectively.

The maximum of n_i for a given h is more difficult. This upper bound is known for benzenoids (Gutman 1982). It is inferred that an extremal coronoid with $n_i = (n_i)_{max}$ can be deduced from the corresponding extremal benzenoid by introducing a *naphthalene* hole. This makes n_i to decrease by 10, the smallest possible number, which is achieved for this particular corona hole. Any other hole results in a larger decrease in n_i, which never can be compensated by the larger decrease in h. For a benzenoid with H hexagons and N_i internal vertices the extremal value

$$(N_i)_{max} = 2H + 1 - \{\sqrt{12H - 3}\}$$ (3.12)

is known (Harary and Harborth 1976; Gutman 1982). The symbol $\{x\}$ is to be understood as the smallest integer larger than or equal to x. According to the above reasoning the substitution $H = h+2$ and $(N_i)_{max} = (n_i)_{max} + 10$ imply the maximum of n_i for coronoids at a given h. We have arrived at the net result that a coronoid can be constructed if and only if

$$0 \leq n_i \leq 2h - 5 - \{\sqrt{12h + 21}\}$$ (3.13)

For a given h, the minimum values $n = n_{min}$, $m = m_{min}$ and $n_b = (n_b)_{min}$ are obviously realized simultaneously in a coronoid in which $n_i = (n_i)_{max}$. These extremal values could be deduced from the corresponding ones for benzenoids, but the result is obtained easier simply be inserting $n_i = (n_i)_{max}$ into eqns. (1), (2) and (5). It was arrived at:

$$2h + 5 + \{\sqrt{12h + 21}\} \leq n \leq 4h$$ (3.14)

$$3h + 5 + \{\sqrt{12h + 21}\} \leq m \leq 5h$$ (3.15)

$$10 + 2\{\sqrt{12h + 21}\} \leq n_b \leq 4h$$ (3.16)

The possible numbers of vertices in coronoids are $n = 32$ and $n \geq 35$ (not 33 or 34). For the number of edges: $m = 40$, 44, 45 and $m \geq 47$ (not 41, 42, 43 or 46). The numbers of boundary vertices are clearly even; the possible values are $n_b = 32$, 34, 36, 38, ...

Since the corona hole may be interpreted as a benzenoid, any relation for benzenoids — and especially anyone for extremal values of invariants — may be transferred to corona holes. Here we give two such relations.

$$0 \leq n_i^o \leq 2h^o + 1 - \{\sqrt{12h^o - 3}\}$$ (3.17)

$$2\{\sqrt{12h^o - 3}\} \leq n_b'' \leq 4h^o + 2$$ (3.18)

Finally in this paragraph we give a special formula for coronoids, viz.

$$h \geq h^o + 3 + \{\sqrt{12h^o - 3}\} \tag{3.19}$$

which gives the smallest number of hexagons (h) for a coronoid when the size of the hole (h^o) is given.

3.2.4 *Coloring of Vertices. More Invariants*

The vertices of a coronoid system may be colored black and white just as in a benzenoid system, so that no adjacent vertices have the same color.

Convention: Throughout this book, when vertices are indicated by circles, they are always colored. Furthermore, peaks are always colored white, whereby the valleys become black.

The color excess, Δ, is another invariant defined (for coronoids as well as benzenoids) by

$$\Delta = |n^{(b)} - n^{(w)}| \tag{3.20}$$

where $n^{(b)}$ and $n^{(w)}$ denote the numbers of black and white vertices, respectively. One has also

$$\Delta = |n_i^{(b)} - n_i^{(w)}| \tag{3.21}$$

where $n_i^{(b)}$ and $n_i^{(w)}$ are the numbers of black and white internal vertices, respectively. Finally

$$\Delta = |n_v - n_\wedge| \tag{3.22}$$

Figure 3 shows a coronoid system with $\Delta=1$ in two orientations. The example de-

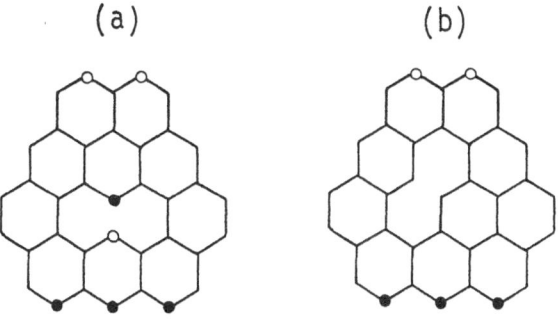

Fig. 3.3. A coronoid with $\Delta=1$. The numbers of valleys and peaks are: (a) $n_v = 4$, $n_\wedge = 3$; (b) $n_v = 3$, $n_\wedge = 2$.

monstrates that the numbers of peaks and valleys ($n_\wedge$ and $n_\vee$) may depend on the particular orientation, but the invariant Δ is a unique integer (or zero) for a given coronoid.

For coronoid (like benzenoid) systems all values $\Delta = 0, 1, 2, 3, \ldots$ are possible.

The following propositions are fairly obvious; they hold for coronoids as well as for benzenoids. If $n_i = 0$, then $\Delta = 0$. If n_i is an even number, then Δ may only assume an even number or zero. If n_i is odd, then Δ must be odd. The lower and upper bounds for Δ are given by

$$\frac{1}{2}[\, 1 - (-1)^{n_i}\,] \leq \Delta \leq n_i \tag{3.23}$$

All Δ values within the specified restrictions may be realized. Examples: for $n_i = 8$, $\Delta = 8, 6, 4, 2$ or 0; for $n_i = 7$, $\Delta = 7, 5, 3$ or 1.

3.3 CLASSIFICATIONS

3.3.1 *Introduction*

Many principles of classification of coronoids and benzenoids have been put forward. Here we concentrate upon a few of the fundamental principles for coronoids. First a classification according to the number of internal vertices is established. Next several classifications with relevance to Kekulé structure counts are treated in some detail. Finally the symmetry groups are invoked and subdivisions within some of them are specified.

3.3.2 *Classification According to Internal Vertices*

A *catacondensed coronoid* is a coronoid with no internal vertices ($n_i = 0$). Otherwise, if $n_i > 0$, the coronoid is a *pericondensed coronoid*.

If a catacondensed coronoid is unbranched, it possesses the hexagon modes L_2 and A_2 (where L_2 may be absent). An unbranched catacondensed coronoid is actually synonymous with a primitive coronoid (cf. Paragraph 2.1.1). A catacondensed coronoid is branched if the mode A_3 is present; Fig. 4 provides an example. In a branched catacondensed coronoid the L_1 and A_2 modes are necessarily present too, while L_2 again may be present or absent.

A pericondensed coronoid possesses at least one hexagon whose mode differs from L_1, L_2, A_2 and A_3. The pericondensed coronoids may be divided further according to the number of internal vertices.

Figure 5 shows in the top row a coronoid system with $n_i = 0$ (catacondensed)

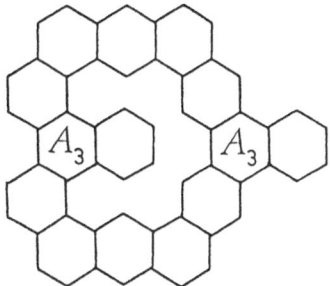

Fig. 3.4. A branched
catacondensed coronoid.
$K = 737$.

and one each with n_i = 1, 2 and 3 (all pericondensed). Another system with n_i = 3 is
depicted at the bottom.

3.3.3 Kekuléan and non-Kekuléan Coronoids

A coronoid, like a benzenoid, may possess Kekulé structures ($K > 0$) or not
(K=0). This gives raise to the classification into Kekuléan coronoids ($K > 0$) and
non-Kekuléan coronoids (K=0). Kekuléan/non-Kekuléan coronoids (or benzenoids) are
also simply called Kekuléans/non-Kekuléans.

All Kekuléan coronoids have Δ=0, but this is not a sufficient condition for a
coronoid to be Kekuléan. Non-Kekuléans with Δ=0 exist and are referred to as concealed non-Kekuléans (K=0, Δ=0). All other non-Kekuléans (K=0, $\Delta > 0$) are called
obvious non-Kekuléans. Four obvious non-Kekuléan coronoid systems are found in Fig.
5. Figure 6 shows two concealed non-Kekuléan coronoids.

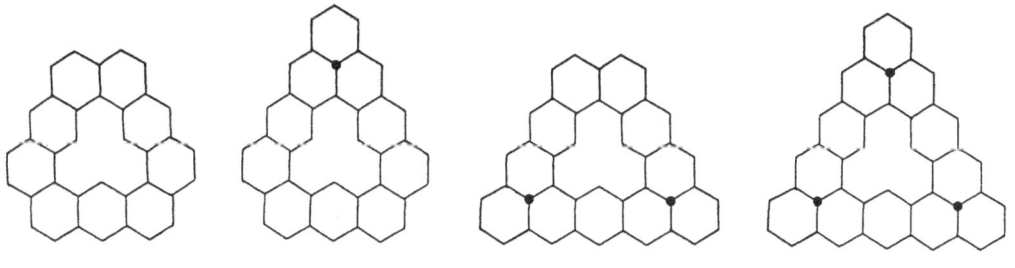

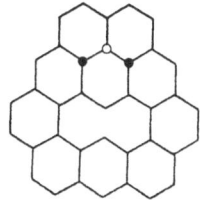

Fig. 3.5. Coronoids with n_i = 0, 1, 2, 3 in the top row; n_i = 3 for the bottom system. Values of color excess: Δ = 0, 1, 2, 3 in the top row; Δ=1 (see also Fig. 3.3)
for the bottom system.

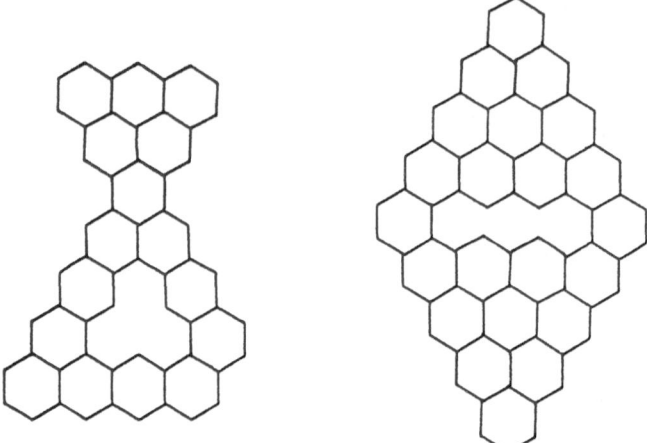

Fig. 3.6. Two concealed non-Kekuléan (K=0, Δ=0) coronoid systems.

Convention: Throughout this book, obvious non-Kekuléan coronoid systems are always drawn so that $n_V > n_\Lambda$.

3.3.4 *The "neo" Classification*

The *neo* classification is applicable to coronoid- as well as benzenoid systems. Within this classification the Kekuléan systems are subdivided into two classes. A coronoid system is essentially disconnected if it has fixed bonds, i.e. double and/or single in the same positions of all Kekulé structures. If it has no fixed bonds the coronoid system is called normal.

All catacondensed coronoids are normal. Figure 7 shows two pericondensed normal coronoids. All essentially disconnected coronoids are pericondensed, as also is the case with essentially disconnected benzenoids. Also in both cases a junction

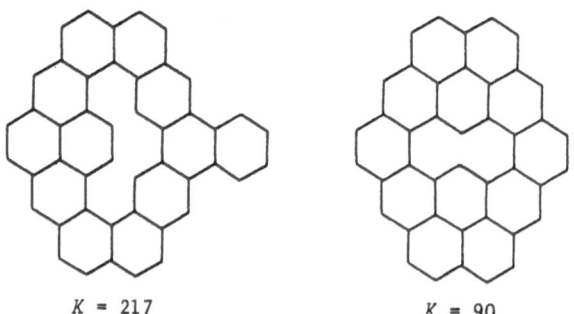

K = 217 K = 90

Fig. 3.7. Two normal (Kekuléan) pericondensed coronoid systems. K numbers are given.

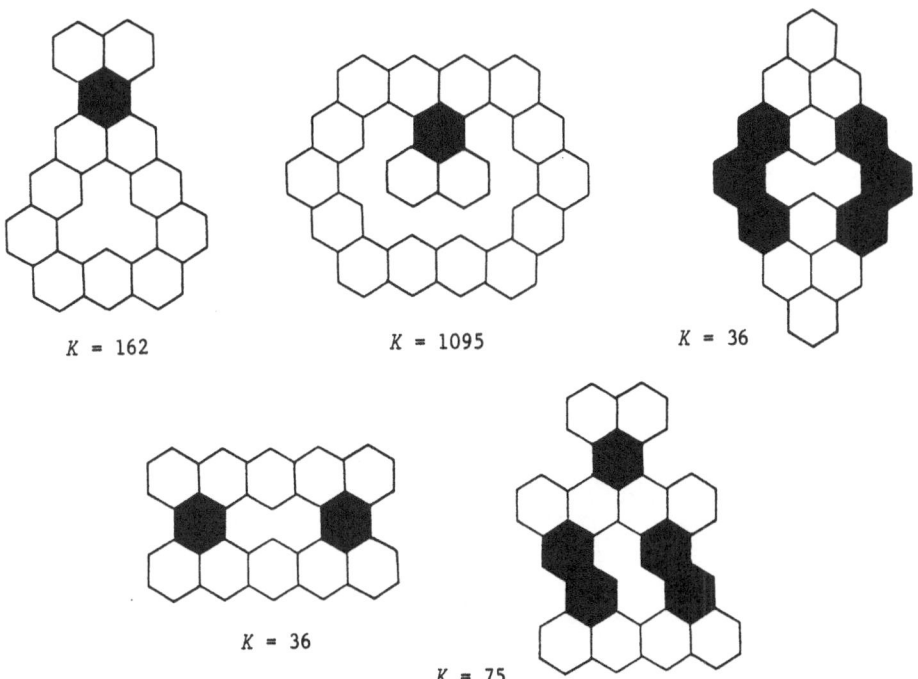

Fig. 3.8. Five essentially disconnected (Kekuléan, pericondensed) coronoid systems. The junctions are black. K numbers are given.

and effective units are recognized (Brunvoll, Cyvin BN, Cyvin and Gutman 1988a). In Fig. 8 five essentially disconnected coronoids are depicted.

The *neo* classification takes into account all coronoid systems. They may be either normal (n), essentially disconnected (e) or non-Kekuléan (o).

Normal additions of a hexagon are defined as the one-, three- and five-contact additions (cf. Fig. 2.2), i.e. those additions where the added hexagon acquires the mode L_1, L_3 or L_5 (cf. Fig. 1). It has been conjectured (Cyvin and Gutman 1986; Cyvin SJ, Brunvoll J, Cyvin BN 1989b) that all normal benzenoids with $h+1$ hexagons can be generated from the set of normal benzenoids with h hexagons by normal additions. This conjecture has been supported by a vast amount of computerized enumerations. Finally the proposition was also proved mathematically (He WC and He 1990a; 1990b), and thus the status of the conjecture was raised to a theorem. This theorem provides an alternative definition of normal benzenoids. It also implies that it must be possible to subject any normal benzenoid to a so-called normal tearing down, hexagon by hexagon, right down to one hexagon (*benzene*). In this process only a hexagon of the mode L_1, L_3 or L_5 should be removed every time.

The situation for coronoids is not so simple as described above. It can be conjectured with good confidence that normal additions to normal coronoids create new normal coronoids only. The left-hand system of Fig. 7 can be obtained in this way. However, there exist normal coronoids which can not be generated from normal coronoids by normal additions. The right-hand system in Fig. 7 is an example. We conclude: it is not possible to generate all normal coronoids by normal additions to the primitive coronoids.

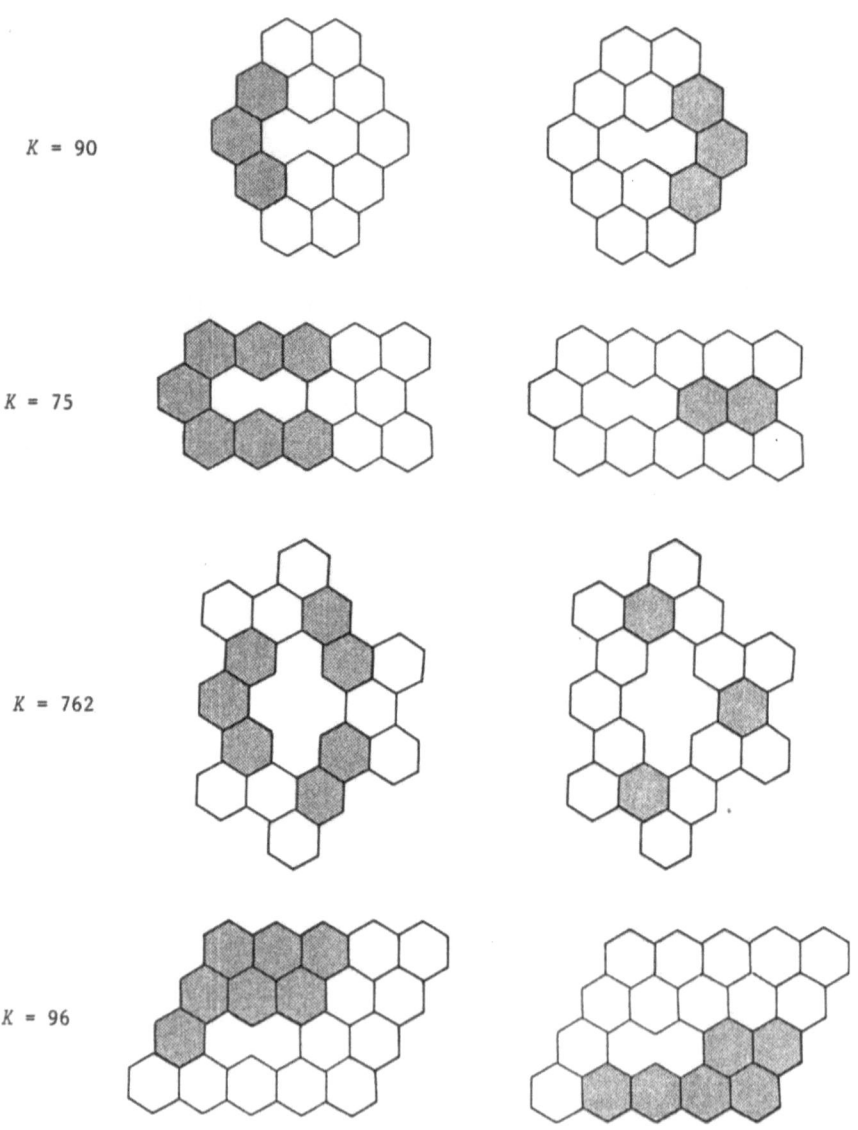

$K = 90$

$K = 75$

$K = 762$

$K = 96$

Fig. 3.9. Four half essentially disconnected (HED) coronoids. The two schemes for Kekulé structures are indicated. K numbers are given.

3.3.5 *Introduction to Regular and Half Essentially Disconnected Coronoids*

Extensive studies of the Kekulé structures of coronoids demonstrated the need for a subdivision of the normal coronoid systems. This led to the concepts *regular* and *half essentially disconnected* (HED) *coronoids*.

As an introduction to the new categories of coronoids we point out that Fig. 7 shows one example of each: a regular coronoid to the left and an HED coronoid to the right.

Half essentially disconnected coronoids are characterized by the presence of two sets of Kekulé structures, each having fixed bonds. The two sets together should give the complete set of Kekulé structures. Figure 9 shows the two schemes of Kekulé structures for four HED coronoids, including the one of Fig. 7. In each of the two schemes junctions (grey in Fig. 9) and effective units (white) are recognized similarly as in essentially disconnected systems. In one scheme, however, in contrast to an essentially disconnected system, there may be one effective unit only.

As a supplementary example we show a "mixed" system in Fig. 10. It is essentially disconnected with two effective units, *naphthalene* and an HED coronoid. The whole system does possess fixed bonds and is therefore essentially disconnected in accord with the definitions.

The above characterization of HED coronoids in terms of two schemes of Kekulé structures does not serve as a strict definition of this class. It is attempted to give an indirect definition of HED coronoids in the following via a definition of the class of regular coronoids.

3.3.6 *Regular Addition and Regular Tearing Down*

We define the *regular additions* as the normal additions (pertaining to L_1, L_3 and L_5; see above) plus the two types of corona-condensations (cf. Paragraph 2.1.1), which may be linear or angular. The added hexagon may acquire the mode L_2 or A_2, respectively. The opposite process of a regular addition is a *regular tearing down*

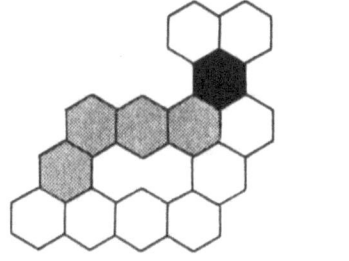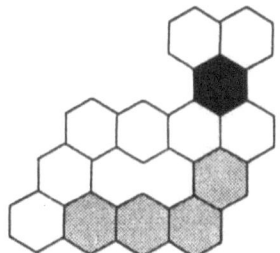

Fig. 3.10. An essentially disconnected coronoid where one of the effective units is half essentially disconnected. $K = 78$.

or in other words a successive removing of hexagons of the modes L_1, L_2, A_2, L_3, L_5 (L_2 or A_2 removed only once for single coronoids).

By definition, the regular additions applied successively (starting from *benzene*) produce exactly all normal benzenoids and regular coronoids. In consequence, any regular coronoid may be subjected to a regular tearing down, hexagon by hexagon, right down to *benzene*. It is not arbitrary in which sequence the hexagons are torn down if the process shall meet with success to its ultimate end. Figure 11 is supposed to shed some light into this problem.

It is clear that a regular tearing down may be considered as successful if one arrives at a normal benzenoid (exemplified by the left-hand drawing of Fig. 11) or if a normal tearing down results in a primitive coronoid (see Fig. 11, middle drawing). Primitive coronoids are all regular and may be torn down regularly in an obvious way. However, if the two hexagons of the right-hand drawing of Fig. 11 are removed as indicated, the resulting coronoid can not be torn down regularly to the end; it is an HED coronoid.

The last statement of the preceding paragraph is evidently consistent with saying that there exist coronoids which can not be torn down to a primitive coronoid by a normal tearing down. It might be tempting to infer that this property distinguishes between coronoids which are regular or not. This, however, would be an over-simplification. As a matter of fact there exist regular coronoids which can not be generated by normal additions to a primitive coronoid and therefore neither torn down normally to a primitive coronoid. The smallest example is shown in Fig. 12. Here it is clear that the system is a regular coronoid, inasmuch as the removing of an L_2 or A_2 hexagon (a regular tearing down) immediately brings it over to a normal benzenoid.

3.3.7 *The "rheo" and "rio" Classifications*

It is conjectured that normal coronoids which are not regular, are half essentially disconnected. They should possess the characteristics of having two schemes

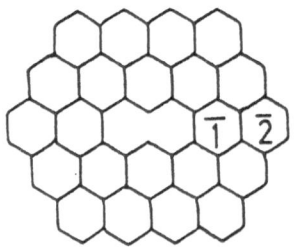 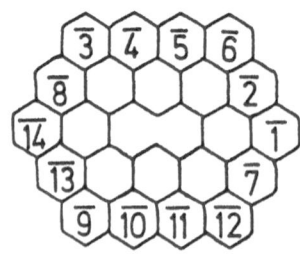 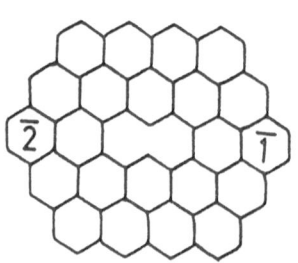

Fig. 3.11. Examples of the regular tearing down of a regular coronoid. Left and middle: correct; right: incorrect. The hexagons should be removed in the sequence $\bar{1}$, $\bar{2}$, $\bar{3}$, ...

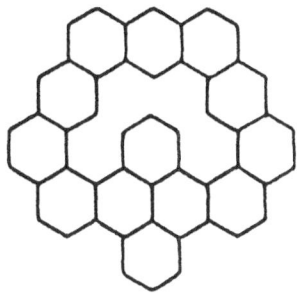

Fig. 3.12. The smallest (h = 13) regular coronoid which can not be generated by normal additions to a primitive coronoid. K = 364.

of Kekulé structures as described in Paragraph 3.3.5. No counterexample to this conjecture has been found in spite of extensive analyses of computer-generated coronoids.

The homolog series of coronoids in Fig. 13 is an illuminating example. By removing one of the L_2 or A_2 hexagons in each system (i.e. by opening the cycle), one obtains from left to right: an essentially disconnected benzenoid, a concealed non-Kekuléan benzenoid, and another concealed non-Kekuléan benzenoid.

The *rheo* classification of coronoids refers to the regular (r), half essentially disconnected or HED (he), essentially disconnected (e) and non-Kekuléan (o) systems. In accord with the above discussion (and provided that the conjectures are valid) these categories cover all (single) coronoids.

Kekuléan coronoids which are not regular, are referred to as *irregular*. Accordingly, they consist of the HED and essentially disconnected systems.

The *rio* classification is coarser than the *rheo* classification. It refers to the categories regular (r), irregular (i) and non-Kekuléan (o).

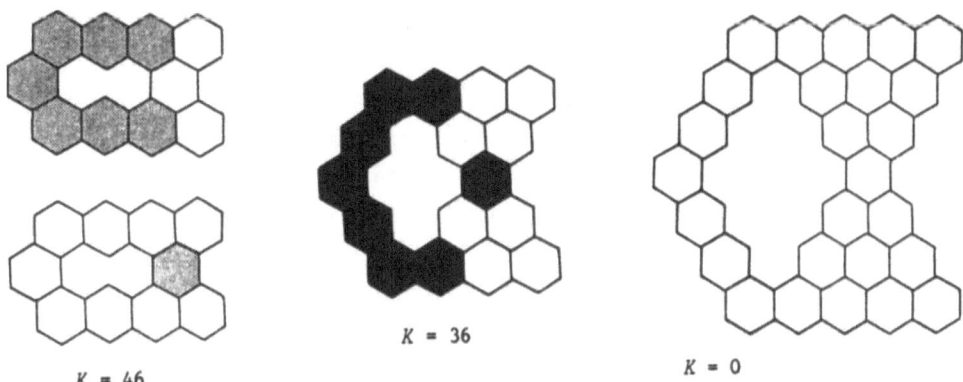

K = 46

K = 36

K = 0

Fig. 3.13. A half essentially disconnected (HED) coronoid (left), essentially disconnected (middle) and concealed non-Kekuléan (right). K numbers are given.

Below we give a survey of the different categories of coronoids defined in
Paragraphs 3.3.3 - 3.3.7.

$$
\text{Kekuléan}
\begin{cases}
\text{normal } (n)
\begin{cases}
\text{regular } (r) \\[1em]
\begin{array}{l}\text{half essentially}\\ \text{disconnected } (he)\end{array} \left.\vphantom{\begin{array}{c}a\\b\\c\\d\end{array}}\right\} \text{irregular } (i)\\[1em]
\begin{array}{l}\text{essentially}\\ \text{disconnected } (e)\end{array}
\end{cases}\\[3em]
\text{non-Kekuléan } (o)
\begin{cases}
\text{obvious} \\[0.5em]
\text{concealed}
\end{cases}
\end{cases}
$$

3.3.8 *Symmetry*

A coronoid (like a benzenoid) may belong to the symmetry group D_{6h} or one of
its subgroups which include reflection in the horizontal plane: C_{6h}, D_{3h}, C_{3h}, D_{2h},
C_{2h}, C_{2v}, C_{s}.

The corona hole, which is to be interpreted as a benzenoid, may also have one
of the eight mentioned symmetries, but not necessarily the same as the coronoid itself. The coronoid has either the same symmetry as the corona hole, say G^{o}, or a
lower symmetry in the sense of a subgroup of G^{o}. The scheme of subgroups is given
in Fig. 14. It should be understood so that every intermediate step can be skipped.
Thus C_{s} is a subgroup of all of the other groups.

For most of the groups under consideration, actually all but C_{2v} and C_{s}, the
system has a unique point which remains unaffected by all symmetry operations and
is called the *centre*. (For C_{2v} the points on a unique symmetry axis are unaffected
by the symmetry operations, while absolutely all points are unaffected in the case
of C_{s}.)

Assume that a coronoid belongs to one of the symmetries D_{6h}, C_{6h}, D_{3h}, C_{3h},
D_{2h} or C_{2h} and consequently has a centre. This centre must inevitably fall in the
centre of the corona hole (which also must have one of the six mentioned symmetries,
either the same as the coronoid or higher). In the following descriptions remember
that the corona hole is interpreted as a benzenoid. It is also important to adher to
the convention that each hexagon has two vertical edges. Only single coronoids (one
hole) are taken into account.

<u>1</u>. A coronoid of *hexagonal* symmetry belongs by definition to the symmetry
group D_{6h} (*regular hexagonal*) or C_{6h}.

The centre of a coronoid of hexagonal symmetry falls in a centre of a hexagon
(of the corona hole). Hence we may speak about a *central hexagon*.

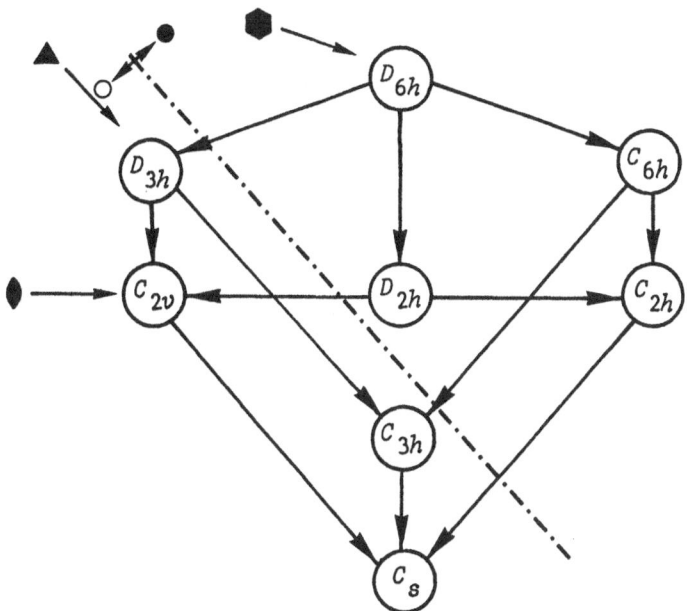

Fig. 3.14. The hierarchy of the symmetry group D_{6h} and its subgroups with relevance
to benzenoids and coronoids. A subgroup of G is found below or to the side of G in
the direction of the arrows. The presence of principal (highest) symmetry axes are
indicated: ⬢ six-fold; ▲ three-fold; ◆ two-fold. Centre of symmetry (●) is found
for the groups on one side of the broken line as indicated (○ indicates absence of
a centre of symmetry).

2. A coronoid of *trigonal* symmetry may belong to D_{3h} *(regular trigonal)* or
C_{3h}. There are two kinds, viz.:

(i) Coronoids of trigonal symmetry with a central hexagon.

(ii) Coronoids of trigonal symmetry where the centre coincides with a vertex;
they are said to have a *central vertex*.

The regular trigonal (D_{3h}) systems of the first kind (i) are subdivided into:

(a) those where the two-fold symmetry axes do not pass through any vertex, but
bisect edges perpendicularly, and

(b) those where the two-fold symmetry axes pass through vertices and edges.

The D_{3h} systems of the second kind (ii) all have two-fold symmetry axes passing through vertices and edges as in (b).

The cases (a) and (b) may alternatively be characterized by saying that the
systems have (a) a horizontal two-fold symmetry axis or (b) a vertical two-fold symmetry axis, respectively. Illustrations:

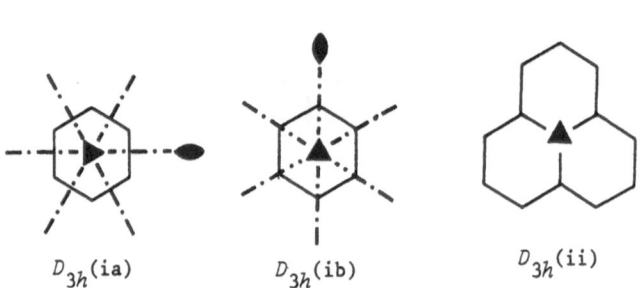

D_{3h}(ia) D_{3h}(ib) D_{3h}(ii)

3. A coronoid of *dihedral* symmetry (D_{2h}) and a *centrosymmetrical* (C_{2h}) coronoid may belong to one of the two kinds:

(i) The system has a central hexagon.

(ii) The centre bisects one edge; this edge is referred to as the *central edge*.
Illustrations:

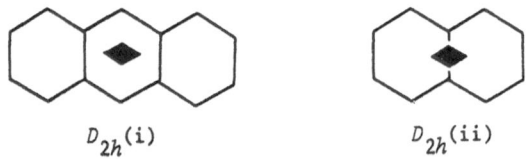

D_{2h}(i) D_{2h}(ii)

4. A coronoid system of C_{2v} is called *mirror-symmetrical*. These systems are subdivided into:

(a) those where the two-fold symmetry axis does not pass through any vertex, but bisects edges perpendicularly, and

(b) those where the two-fold symmetry axis passes through vertices and edges.

The C_{2v} systems can (within the usual conventions) be oriented so that the (unique, two-fold) symmetry axis is either (a) horizontal or (b) vertical.

5. A coronoid belonging to C_s is called *unsymmetrical*.

In conclusion of this paragraph we give a schematic survey of the symmetry groups and their subdivisions.

D_{6h} ⎫
C_{6h} ⎬ central hexagon

D_{3h} {
  (i) central hexagon { (a) 2-fold axes bisect edges
                                (b) 2-fold axes through edges
  (ii) central vertex
}

C_{3h} {
  (i) central hexagon
  (ii) central vertex
}

D_{2h}, C_{2h} {
  (i) central hexagon
  (ii) central edge
}

C_{2v} {
  (a) 2-fold axis bisects edges
  (b) 2-fold axis through edges
}

C_s

3.4 AROMATIC SEXTETS AND ALL-CORONOIDS

3.4.1 *Aromatic Sextet*

An aromatic sextet (Clar 1964; 1972) is defined, in connection with a Kekulé structure, by a hexagon whose edges correspond to three double (and three single) bonds (as in *benzene*). Sometimes it is distinguished between proper and improper sextets (Ohkami et al. 1981), chosen (arbitrarily) as having the vertical double bond at the right-hand or left-hand side, respectively:

  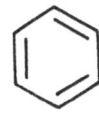

                                proper sextet   improper sextet

An aromatic sextet is conventionally represented by a circle inscribed in the hexagon (see above).

3.4.2 *Clar Structure*

A Clar structure (Clar 1972) of a coronoid (or benzenoid) represents a bonding scheme with a maximum number of aromatic sextets. The Clar structures are constructed according to the following formal rules (Gutman 1982): (a) circles are never drawn in adjacent hexagons; (b) the remainder of the system, obtained by deletion of the vertices of the hexagons that possess circles, must have a Kekulé structure; this remainder may be empty (in the case of all-benzenoids and all-coronoids); (c) as many circles as possible are drawn, subject to the constraints (a) and (b).

The Clar structure of *kekulene* has been given several times (Aihara 1976; Krieger et al. 1979; Balaban 1982; Staab, Diederich, Krieger and Schweitzer 1983; Staab, Diederich and Čaplar 1983; Vogler 1985; Funhoff and Staab 1986; Cyvin SJ, Brunvoll, Cyvin and Brendsdal 1988). It is shown in Fig. 15 (right-hand side); see also Fig. 1.2. The left-hand side of Fig. 15 shows the Clar structure of the smallest (primitive) coronoid, viz. *cyclo*[d.e.e.e.d.e.e.e]*octakisbenzene.*

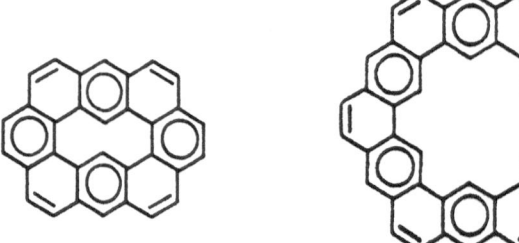

Fig. 3.15. The Clar structures of *cyclo*[d.e.e.e.d.e.e.e]*octakisbenzene* (left) and *kekulene* (right).

The Clar structures for *cyclo*[d.e.e.e.d.e.e.e]*octakisbenzene* and for *kekulene* are unique, but this is not so in general for a coronoid (or benzenoid). It has been pointed out, for instance, that *cyclo*[d.e.d.e.e.d.e.d.e.e]*decakisbenzene* has nine Clar structures (Staab, Diederich and Čaplar 1983), each holding four aromatic sextets. They have been depicted by Brendsdal et al. (1988); see also Fig. 16.

Any aromatic sextet in a Kekulé structure can be reversed (from proper to improper and vice versa) to produce another Kekulé structure. Thus a Clar structure represents a number of Kekulé structures, which is 2^{α} if α denotes the number of aromatic sextets. One of the $2^{6} = 64$ Kekulé structures "contained" in the Clar structure of *kekulene* is found as the right-hand drawing of Fig. 2.3. It should also be mentioned that the situation may be reverse: it may happen that one and the same Kekulé structure is contained in different Clar structures.

In a generalized Clar structure only the proper sextets are indicated (by circles). The double bonds which do not belong to the proper sextets are drawn as

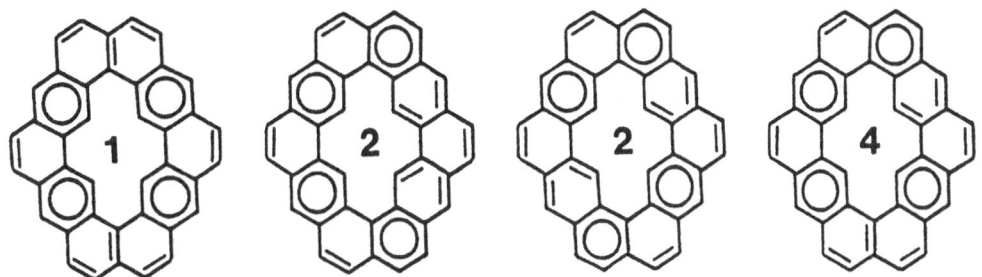

Fig. 3.16. Clar structures for *cyclo*[d.e.d.e.e.d.e.d.e.e]*decakisbenzene*. The inscribed numbers are multiplicities (for symmetrically equivalent structures).

ordinary edges. It is allowed for less than the maximum number of proper sextets, including the trivial generalized Clar structure with no proper sextets. But also every Clar structure corresponds to a generalized Clar structure.

3.4.3 *All-Coronoid*

The definition of an all-coronoid was foreshadowed in the preceding paragraph. It is analogous to the definition of all-benzenoids, systems which also are called fully benzenoids.

An *all-coronoid* system (or simply all-coronoid) has Kekulé structures where all the double bonds belong to aromatic sextets. The definition in the wording of Randić (1980) is applicable to all-coronoids as well as all-benzenoids; it reads: "For these compounds one can write a valence structure in which a ring is either represented as an isolated sextet or is devoid of conjugation." Here a valence structure is synonymous with what we call a Kekulé structure.

To be more precise, it is possible in an all-coronoid (or all-benzenoid) to assign uniquely a constellation of aromatic sextets throughout the system, so that the remaining hexagons do not possess additional double bonds in the appropriate Kekulé structures. The sextets determine what is called the full hexagons (F), while the other hexagons are called empty (E). It is usual to inscribe circles in the full hexagons. Then the picture is precisely a manifestation of the Clar structure, which is unique for any all-coronoid (and all-benzenoid). Let the number of full and empty hexagons be denoted by h_F and h_E, respectively; then $h = h_F + h_E$. Evidently the Clar structure has h_F aromatic sextets, and their unique constellation is found in 2^{h_F} Kekulé structures.

The very term "all-coronoid" was coined by Cyvin SJ, Bergan and Cyvin (1987), but the existence of all-coronoids was already pointed out ten years before by Polansky and Rouvray (1977). The smallest such system (see Fig. 17, the extreme left-hand drawing) was identified by Bergan et al. (1987), and more examples followed (Cyvin SJ, Bergan and Cyvin 1987; Cyvin SJ, Brunvoll, Cyvin and Brendsdal

1988; Brendsdal et al. 1988; Gutman and Cyvin 1989b). Many of the topological properties for all-benzenoids, which have been derived (Polansky and Rouvray 1976b;
Polansky and Gutman 1980; Gutman and Cyvin 1988), are immediately applicable to all-
coronoids.

Any all-coronoid (as well as all-benzenoid) is normal (Kekuléan). Furthermore,
any all-coronoid is regular; cf. Section 3.3 (especially Paragraph 3.3.5 and the
scheme at the end of Paragraph 3.3.7) for the different classifications. An all-
coronoid may be either catacondensed or pericondensed (cf. Paragraph 3.3.2). Some
examples are depicted in Fig. 17 (catacondensed) and Fig. 18 (pericondensed).

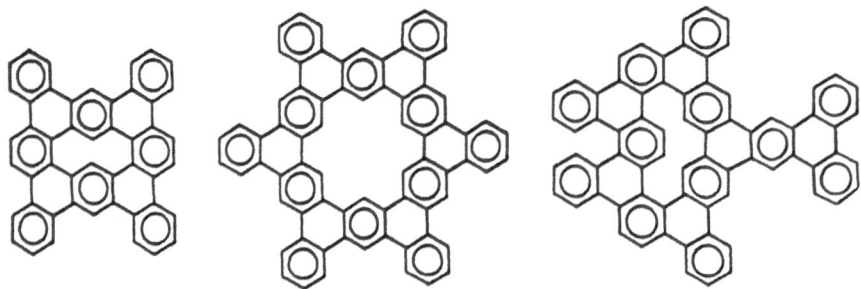

Fig. 3.17. Some catacondensed all-coronoids.

Fig. 3.18. Some pericondensed all-coronoids.

Chapter 4

PRIMITIVE CORONOIDS (I) AND ANNULENES: ANATOMY

4.1 INTRODUCTION

In the preceding chapter the "anatomy" (as characterized in the introduction; Section 3.1) of coronoids in general was treated. In the present chapter we point out some additional topological properties pertaining to the special coronoids called primitive, these basic building stones of coronoid systems. The definition and a treatment of single unbranched chains associated with the primitive coronoids are included. Finally a few properties of *annulenes* among degenerate coronoid systems are summarized.

Some of the material in this chapter has been exposed previously in different places (Cyvin 1988; 1989; Brunvoll, Cyvin BN, Cyvin, Gutman, Tošić and Kovačević 1989; Cyvin SJ, Brunvoll and Gutman 1990).

4.2 SOME TOPOLOGICAL PROPERTIES OF PRIMITIVE CORONOIDS

4.2.1 *Definition*

A primitive coronoid is an unbranched catacondensed coronoid system. In other words it consists of a single unbranched chain of hexagons in a circular (corona-condensed) arrangement; see also Paragraph 3.3.2.

The dualist of a primitive coronoid is a cycle.

Primitive coronoids as a subclass of coronoids are also fully characterized by possessing only linearly and angularly annelated hexagons, which shall be identified by L and A, respectively. To be precise, the modes are actually L_2 and A_2, respectively. A primitive coronoid can be constructed so that it contains A hexagons only (but not L hexagons only).

4.2.2 *Corona Hole*

A benzenoid with a cove or a fjord (cf. Paragraph 3.2.1 for definitions of these concepts) can obviously not serve as a corona hole for a primitive coronoid. Figure 1 (top row) shows the smallest benzenoid with a cove (left) and with a fjord (right) attempted to be used as a corona hole for primitive coronoids. The $h = 12$ system in particular (cf. Fig. 1) contains a part of the unconventional interpretation of *pyrene* as shown in Fig. 1.4.

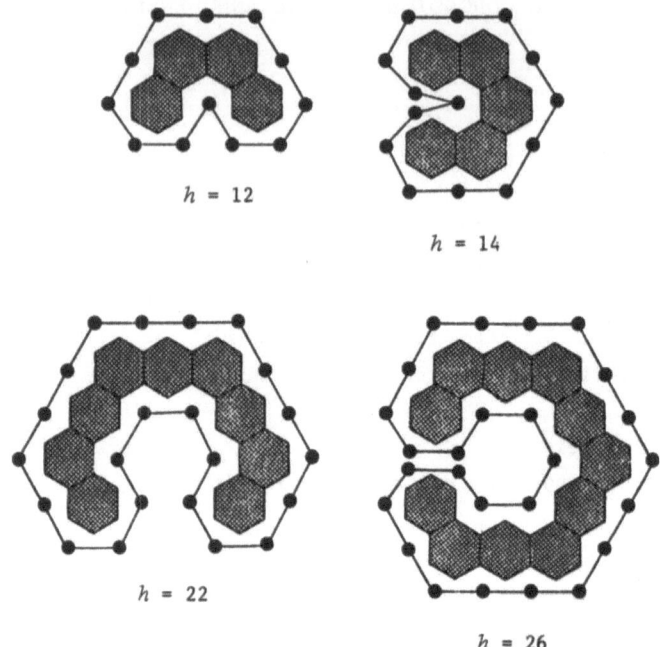

$h = 12$

$h = 14$

$h = 22$

$h = 26$

Fig. 4.1. Four benzenoids (grey hexagons) attempted to be used as corona holes for primitive coronoids. Dualists are (tentatively) employed when the holes are circumscribed.

One should not be misled to believe that any benzenoid with more than one hexagon and without any cove or fjord can serve as a corona hole for a primitive coronoid. The bottom row of Fig. 1 shows two counterexamples. The left-hand ($h = 22$) system is the smallest example which was found by trial and error. To make it clear, it represents presumably the benzenoid with the shortest perimeter (it has nine hexagons), no cove and no fjord, and being unsuitable as corona hole for a primitive coronoid. The two systems obtained by circumscribing the holes of the bottom row in Fig. 1 are helicenic quasi-coronoids.

4.2.3 *Corners*

The A hexagons in a primitive coronoid are also called *corners*. A corner may be protruding or intruding. A *protruding corner*, J, has three edges on the outer perimeter, of which exactly one is a free edge. An *intruding corner*, I, has three edges on the inner perimeter, one of them being a free edge. Exactly one free edge in each corner are the only free edges of a primitive coronoid.

In a primitive coronoid there are at least six protruding corners. One set may always be selected so that their free edges point in the six directions:

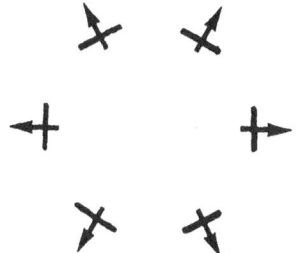

Let the numbers of protruding and intruding corners be denoted by J and I, respectively. Then

$$J = I + 6; \qquad I = 0, 1, 2, \ldots \qquad (4.1)$$

4.2.4 *Segments*

A linear (single) chain of hexagons between two (neighbouring) corners inclusive is called a *segment*. Notice that a corner belongs to two (neighbouring) segments. The number of segments (S) in a primitive coronoid is clearly equal to the number of corners, viz.

$$S = I + J \qquad (4.2)$$

On inserting from (1) it is obtained

$$S = 2I + 6 \qquad (4.3)$$

which shows that the number of segments (or corners) must be an even number. Figure 2 shows that all values $S = 6, 8, 10, 12, \ldots$ are possible. Here the "vertical"

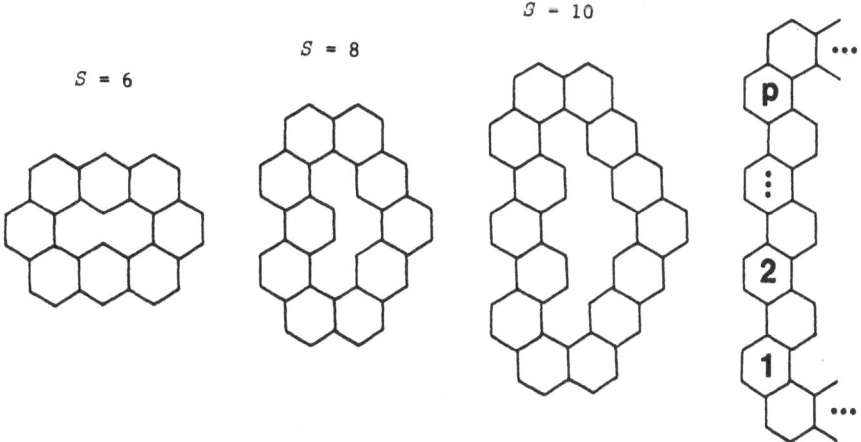

Fig. 4.2. Systematic construction of coronoids with 6, 8, 10, 12, ... segments.

zigzag chain of $2p + 1$ hexagons ($p = 1, 2, 3, \ldots.$) has $2p$ segments. The whole coronoid has 4 segments in addition; hence $S = 2p + 4$.

4.2.5 *Perimeters*

The length of the outer perimeter (i.e. its number of vertices or number of edges) is $n_b{}'$ (cf. Paragraph 3.2.2). Then, for a primitive coronoid,

$$h = \frac{1}{2} n_b{}' - 3 \qquad\qquad (4.4)$$

Similarly, when $n_b{}''$ is the length of the inner perimeter, one has

$$h = \frac{1}{2} n_b{}'' + 3 \qquad\qquad (4.5)$$

Consequently, for a primitive coronoid,

$$n_{b.}{}' = n_b{}'' + 12 \qquad\qquad (4.6)$$

In accord with the statements about perimeter cycles in Paragraph 3.2.2 the outer and inner perimeters of a primitive coronoid are always either both of the $[4k + 2]$ type or both $[4k]$.

4.2.6 *Symmetry*

The symmetry group of a primitive coronoid is the same as the symmetry group of its corona hole. All the eight symmetry groups described in Paragraph 3.3.8 are possible, and also the different subdivisions of some of these groups.

4.3 NOTATION

Let $l_1, l_2, \ldots., l_S$ denote the *lengths* of the segments in a primitive coronoid in terms of the numbers of hexagons. Assume that (l_i, l_{i+1}) are neighbours for $i = 1, 2, \ldots., S-1$, and also (l_S, l_1) are neighbours. In other words the lengths are written down in the order as they appear around a coronoid. Such a set of integers is called the *sequence of segments*. The number of hexagons (h) is:

$$h = -S + \sum_{i=1}^{S} l_i \qquad\qquad (4.7)$$

A coronoid itself is denoted in terms of the sequence of segments as

$$/l_1, l_2, \ldots., l_S/$$

The left-hand system of Fig. 2, for instance, is identified by $/2, 3, 2, 2, 3, 2/$ or $/2, 3, 2^2, 3, 2/$ in a slightly abbreviated form. Another abbreviation for the same system is $/2, 3, 2/^2$. The most compact form reads $/2^2, 3/^2$.

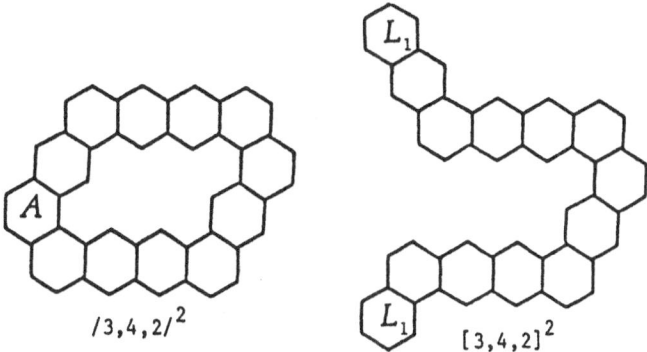

Fig. 4.3. A primitive coronoid and one of the chains associated with it.

It is clear that the same coronoid usually may be denoted in different (equivalent) ways by reversing the lengths and permuting them cyclically. The left-hand system of Fig. 3, for instance, is equal to $/3, 4, 2/^2 = /4, 2, 3/^2 = /2, 3, 4/^2 = /4, 3, 2/^2 = /3, 2, 4/^2 = /2, 4, 3/^2$. On the other hand, a sequence of segments does not always define a primitive coronoid uniquely. The example with the smallest possible coronoids (i.e. with smallest h) occurs for $h = 12$ and is shown below.

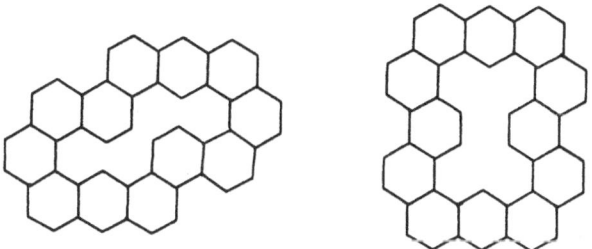

Both of these systems conform with $/2^4, 3/^2$.

4.4 ASSOCIATED CHAIN WITH A PRIMITIVE CORONOID

4.4.1 *Definition*

A single unbranched chain referred to as an *associated chain* with a primitive coronoid is produced in the following way. Choose an (arbitrary) corner (A) and convert it into two end hexagons (L_1); see Fig. 3 for an example. We may also say that the coronoid is associated with the (single unbranched) chain. The way in which the kinks go in an associated chain is immaterial. The associated chain is, by definition, determined by a sequence of segments. We employ a notation analogous with the one for primitive coronoids, but with brackets, viz.

$$[l_1, l_2, \ldots, l_S]$$

The segment lengths of an associated chain can not be permuted in general, only reversed. For the example in Fig. 3 one has $[3, 4, 2]^2 = [2, 4, 3]^2$.

A coronoid has one hexagon less than a chain associated with it.

It is emphasized that different associated chains may in general be produced from one coronoid, depending on which corner is applied for opening of the cycle. The corner may by the way be either protruding or intruding. On the other hand, if a coronoid is associated with a single unbranched chain, then the coronoid is unique as far as its sequence of segments is concerned when reversed and cyclically permuted segments are considered as equivalent.

4.4.2 Single Unbranched Chains

Assume that a single unbranched chain is defined by its sequence of segments. We shall call it an *odd chain* (resp. *even chain*) when it has an odd (resp. even) number of segments. Let the chain be defined in a directed sense so that is has a *beginning* and an *end*. The choice of the beginning (or end) is of course arbitrary. Usually the chain is drawn so that its beginning is at the left-hand side. The directed sense of a chain is important in some of the cases where we in the following define elaborate chains from two units U.

We shall define eight types of single unbranched chains, say V, by combining two identical chains U. Figure 4 should be consulted for examples.

$V = U \times U$ is produced by repeating the sequence of segments in U; in our example $V = [2^3, 3, 2^3, 3] = [2^3, 3]^2$. The system consists of two compressed units U, where one hexagon (grey in Fig. 4) belongs to both units. This hexagon assumes the A mode (actually A_2).

$V = U \times \tilde{U}$ is a similar system, but the sequence of segments in the second unit is reversed; in our example $V = [2^3, 3, 3, 2^3] = [2^3, 3^2, 2^3]$.

$V = U + U$. The two identical units U are compressed so that the shared hexagon assumes the L mode (actually L_2); in our example: $V = [2^3, 3+2-1, 2^2, 3] = [2^3, 4, 2^2, 3]$.

$V = U + \tilde{U}$. Here again the sequence of segments in the second unit is reversed; in our example: $V = [2^3, 3+3-1, 2^3] = [2^3, 5, 2^3]$.

$V = U \searrow U$ is produced by fusing the two units U so that the adjacent hexagons which share the edge of fusion (marked by a heavy line in Fig. 4) are both in the A mode. In this way always a segment of length 2 is found in-between the sequences of segments of the two units. In our example: $V = [2^3, 3, 2, 2^3, 3] = [2^3, 3, 2^4, 3]$.

$V = U \searrow \tilde{U}$ is a similar system, but the sequences of segments in the second unit is reversed; in our example: $V = [2^3, 3, 2, 3, 2^3]$.

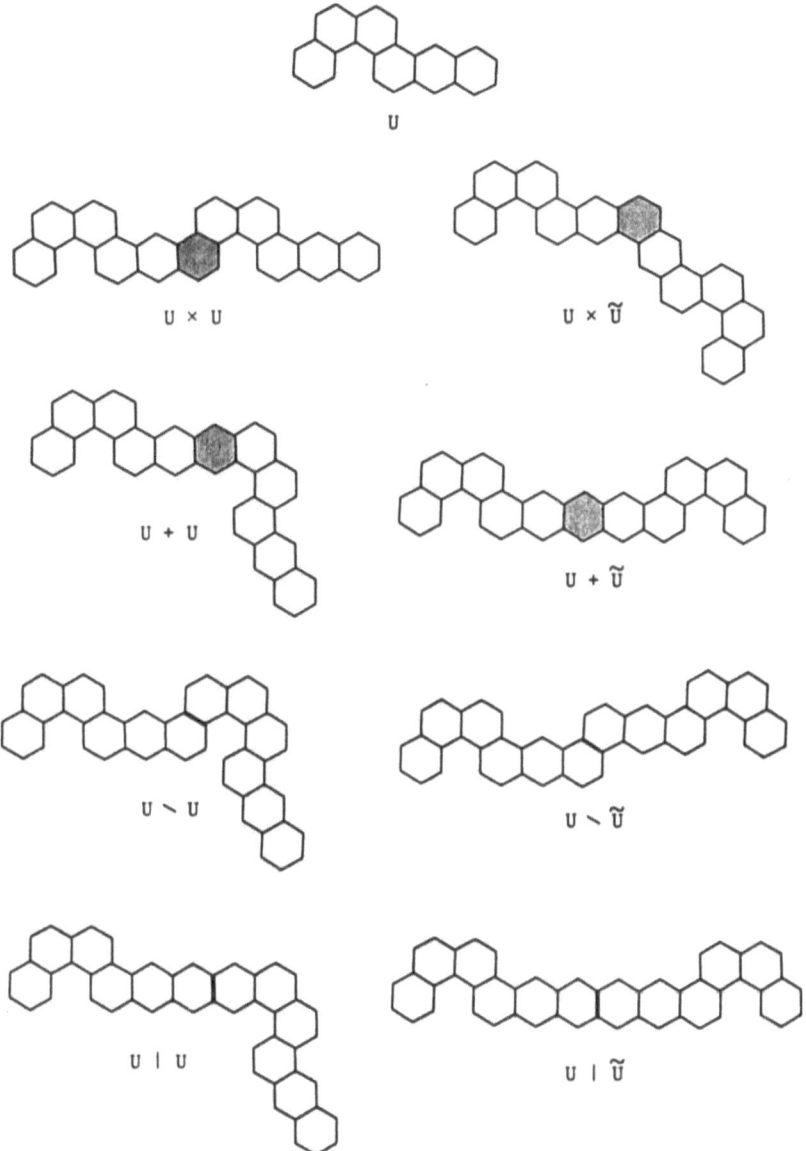

Fig. 4.4. Elaborate chains derived from the chain U.

V = U | U. The two identical units are fused so that the adjacent hexagons which share the edge of fusion are both in the L mode; in our example: V = [2³, 3+2, 2², 3] = [2³, 5, 2², 3].

V = U | Ũ. Here again the sequence of segments in the second unit is reversed; in our example: V = [2³, 3+3, 2³] = [2³, 6, 2³].

Some of these types of chains shall be used as parts of a primitive coronoid in connection with the Kekulé structure count.

4.5 HOLLOW HEXAGON

In accord with the statement of Paragraph 4.2.1 a primitive coronoid has at least six segments. If it has exactly six it is called a *hollow hexagon*. In Fig. 2 the first (left-hand side) system is a hollow hexagon, and so is the coronoid of Fig. 3.

Thus a hollow hexagon is characterized by

$$I = 0, \qquad S = 6 \tag{4.8}$$

A hollow hexagon is completely defined by a sequence of segments. We introduce the notation

$$Hh = /a+1, \; b+1, \; c+1, \; d+1, \; e+1, \; f+1/ \tag{4.9}$$

Figure 5 shows another example. The number of hexagons is

$$h = a + b + c + d + e + f \tag{4.10}$$

Only four of the six parameters are independent by virtue of the connections

$$\tau_1 = a + b = d + e, \qquad \tau_2 = b + c = e + f, \qquad \tau_3 = c + d = f + a \tag{4.11}$$

Let the segments themselves be identified by the symbols a = [a+1], b = [b+1], ..., f = [f+1]. It is clear that the pairs (c, f), (a, d) and (b, e) are mutually parallel. The quantities τ_1, τ_2 and τ_3 are measures of the respective distances between

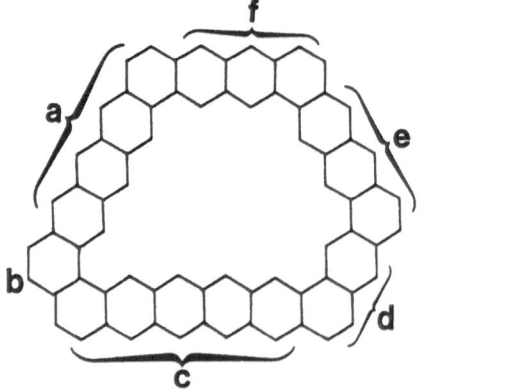
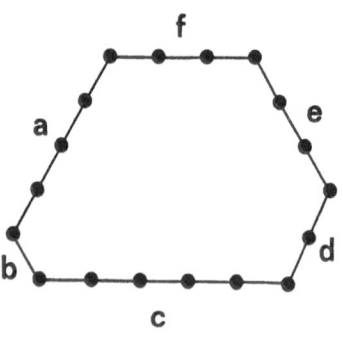

Fig. 4.5. A hollow hexagon ($a=4$, $b=1$, $c=5$, $d=2$, $e=3$, $f=3$, $K = 764$). A representation in terms of the dualist is included (right).

these mutually parallel segments. On eliminating e and f from (10) by means of (11) it is obtained

$$h = a + 2(b + c) + d \qquad (4.12)$$

Figure 5 includes the hollow hexagon represented by a dualist. Notice that each of the parameters (a, b, c, d, e or f) is equal to the number of edges of the dualist for the appropriate segment.

A hollow hexagon may belong to one of the six symmetry groups D_{6h}, D_{3h}, D_{2h}, C_{2h}, C_{2v} and C_s (but not C_{6h} or C_{3h}).

4.6 EXTREMAL PROPERTIES OF THE CORONA HOLES OF PRIMITIVE CORONOIDS

The minimum of h for a coronoid with h^o hexagons of the hole, as given by eqn. (3.19), is realized for a primitive coronoid. But for a primitive coronoid h has also a maximum value when h^o is given. One has altogether

$$h^o + 3 + \{\sqrt{12h^o - 3}\} \le h \le 2h^o + 4 \qquad (4.13)$$

Now we shall consider the opposite problem. For every h value (the number of hexagons of a primitive coronoid) the parameter h^o (number of hexagons of the corona hole) has a minimum and a maximum value, say h_{min}^o and h_{max}^o, respectively.

It has been found (Brunvoll, Cyvin BN, Cyvin, Gutman, Tošić and Kovačević 1989)

$$h_{min}^o = \begin{cases} \dfrac{h-4}{2} \,; & h = 6,\ 8,\ 10,\ 12,\ \ldots . \\[2mm] \dfrac{h-3}{2} \,; & h = 9,\ 11,\ 13,\ 15,\ \ldots . \end{cases} \qquad (4.14)$$

The value $h=6$ pertains to the quasi-coronoid *coronene*.

When h is even the property $h^o = h_{min}^o$ for primitive coronoids is realized exactly for the systems where the corona hole is a catacondensed benzenoid. For a hollow hexagon a catacondensed benzenoid as the corona hole means necessarily that the hole is a single linear chain. This class is characterized by the following sequence of segments.

$$/2,\ 2,\ \frac{h-2}{2}/^2 \,; \qquad h = 6,\ 8,\ 10,\ 12,\ \ldots .$$

Figure 6 shows the primitive coronoids with $h^o = h_{min}^o$ and $h = 12$.

When h is odd the primitive coronoids with $h^o = h_{min}^o$ are realized exactly for benzenoids with one internal vertex as the corona hole. Such a system is never

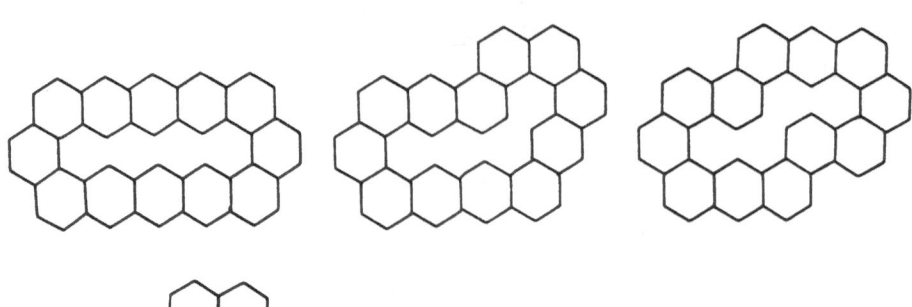

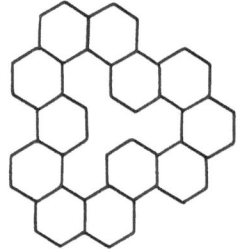

Fig. 4.6. The four primitive coronoids with $h^o = h^o_{min} = 4$ for $h = 12$. The top-left system is a hollow hexagon.

a hollow hexagon when $h > 9$. Figure 7 shows the example for $h = 13$. For hollow hexagons one has

(Hh)
$$h^o_{min} = \begin{cases} \dfrac{h-4}{2} \; ; & h = 6, 8, 10, 12, \ldots. \\[2mm] h-6 \; ; & h = 9, 11, 13, 15, \ldots. \end{cases}$$
(4.15)

instead of eqn. (14). The values for odd h in eqn. (15) are realized in the systems:

$$/2, \; 3, \; \frac{h-5}{2}, \; 3, \; 2, \; \frac{h-3}{2} \; /; \qquad h = 9, 11, 13, 15, \ldots.$$

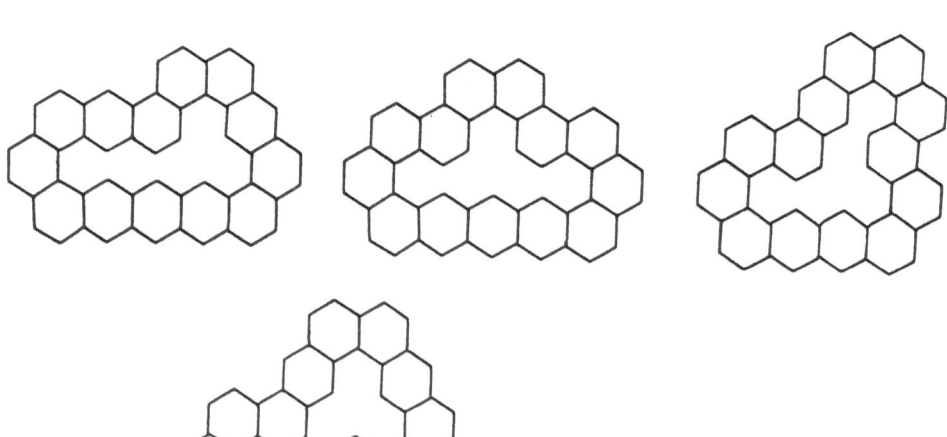

Fig. 4.7. The four primitive coronoids with $h^o = h^o_{min} = 5$ for $h = 13$.

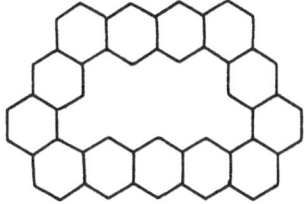

Fig. 4.8. The hollow hexagon with
$h^o = h_{min}^{\;o} = 7$ for $h = 13$.

The example for $h = 13$ is found in Fig. 8.

The maximum values of h^o are not so regular. Based on the analysis of extremal benzenoids by Harary and Harborth 1976 (see also Balaban, Brunvoll et al. 1987) it was obtained (Brunvoll, Cyvin BN, Cyvin, Gutman, Tošić and Kovačević 1989):

$$h_{max}^{\;o} = \left[\frac{h^2}{12} - \frac{h}{2} + 1 \right] \tag{4.16}$$

when $[x]$ is used to designate the largest integer smaller than or equal to x. A set of six equations have been produced, which are equivalent with (16), but give the integers exactly without invoking the bracket symbol. They are strongly related to the Balaban (1971) equations for *annulenes*; cf. the next section. The six equations for $h_{max}^{\;o}(h)$ were compressed into the form:

$$h_{max}^{\;o}(6j + \delta) = \begin{cases} 3j^2 - 3j + 1; & \delta = 0 \\ \\ 3j^2 + (\delta - 3)j; & \delta = 1, 2, 3, 4, 5 \end{cases} \tag{4.17}$$

Here $j = 1, 2, 3, 4, \ldots$, but $j > 1$ for $\delta = 1$.

Some numerical values of $h_{max}^{\;o}(h)$ are entered into Table 1.

The six equations (17) for $\delta = 0, 1, 2, 3, 4, 5$ correspond to characteristic forms of these extremal primitive coronoids. All of them are hollow hexagons. They are specified below (in terms of the lengths of the segments).

$\delta = 0$: $/j+1/^6$

$\delta = 1$: $/j, j+2, (j+1)^3, j+2/$; $j > 1$

$\delta = 2$: $/(j+1)^2, j+2/^2$

$\delta = 3$: $/j+1, j+2/^3$

$\delta = 4$: $/j+1, (j+2)^2/^2$

Table 4.1. Values of h_{max}^{o} for primitive coronoids.

h	h_{max}^{o}	h	h_{max}^{o}
(6)*	1	19	21
8	2	20	24
9	3	21	27
10	4	22	30
11	5	23	33
12	7	24	37
13	8	25	40
14	10	26	44
15	12	27	48
16	14	28	52
17	16	29	56
18	19	30	61

*Quasi-coronoid (*coronene*).

$\delta = 5$: $/j+1, (j+2)^3, j+1, j+3/$

One representative of each form (for $j=2$) is depicted in Fig. 9.

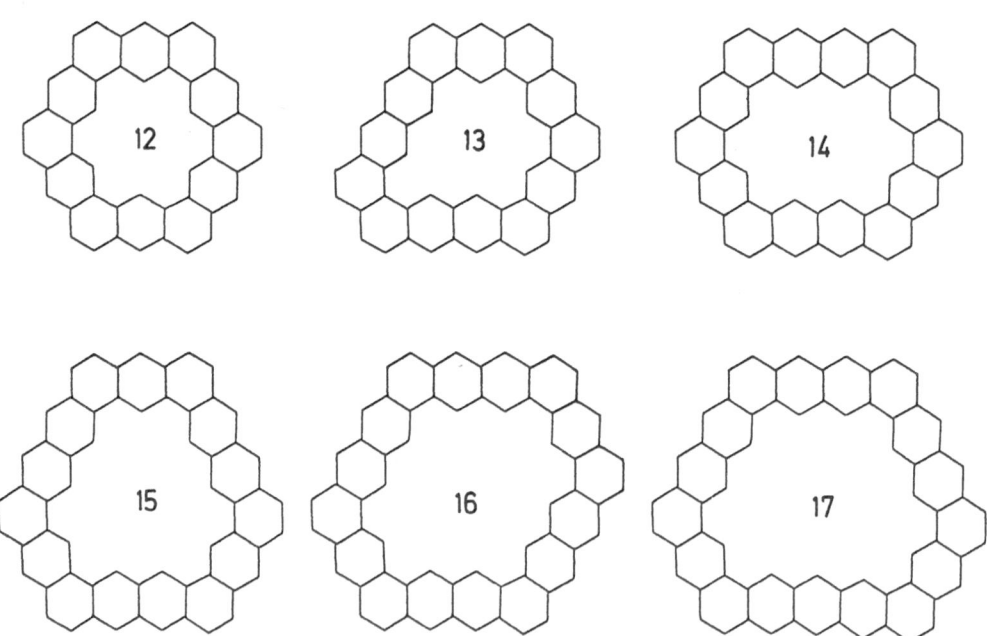

Fig. 4.9. Extremal primitive coronoids with $h^{o} = h_{max}^{o}$ for the inscribed h values. All of them are hollow hexagons.

4.7 ANNULENE

The [*n*]*annulenes* among degenerate coronoid systems are equivalent to cycles on a hexagonal lattice (cf. also Section 2.4). Overlapping edges are not allowed. In the following it is tacitly assumed that the term "*annulene*" refers to an *annulene* in the restricted sense as defined here.

The systems in question have constantly even numbers *n*. They are therefore also referred to as [2*p*]*annulenes*, where

$$n = 2p \qquad\qquad (4.18)$$

More specifically: $n = 6, 10, 12, 14, 16, \ldots$ or $p = 3, 5, 6, 7, 8, \ldots$.

The cycle of the [2*p*]*annulene* may be of the (a) [4*k*+2]-type ($p = 2k + 1$; $k = 1, 2, 3, \ldots$) or (b) [4*k*]-type ($p = 2k$; $k = 3, 4, 5, \ldots$). The benzenoid embraced by these cycles (as perimeters) have an even number of internal vertices (or zero) in the case (a), while case (b) gives an odd number of internal vertices.

There is a one-to-one correspondence between an *annulene* and a benzenoid, since both objects are completely determined by a cycle on a hexagonal lattice. The *annulene* associated with a benzenoid in this way serves as the perimeter of the benzenoid (see also above).

Another correlation will be of special interest for us. Any primitive coronoid may be associated with an *annulene* on identifying the latter with its inner perimeter. *Kekulene*, for instance, is associated with [18]*annulene*. However, the converse is not true; a [2*p*]*annulene* can not serve as inner perimeter of a primitive coronoid when the benzenoid associated with it has a cove or a fjord or otherwise is forbidden as a hole of the coronoid (cf. Paragraph 4.2.2). Still one, more trivial exception exists: *benzene* is recognized as an *annulene* (viz. [6]*annulene*), but does not suit as a corona hole according to the strict definition (Paragraph 2.1.1). *Benzene* alias [6]*annulene* is, however, associated with the "primitive quasi-coronoid" *coronene*.

Many of the topological properties of primitive coronoids have counterparts for *annulenes* through the above correlation. This topic has been treated in detail elsewhere (Cyvin SJ, Brunvoll and Gutman 1990). Some of the aspects are summarized below.

Let h^o denote the number of hexagons associated with a [2*p*]*annulene*. Notice that we utilize the symbol for the number of hexagons of the corona hole of a coronoid; cf. Paragraph 3.2.1. It pertains in particular to the corona hole of the primitive coronoid associated with the *annulene* in question, if this coronoid exists. The basic inter-relation between a primitive coronoid and the *annulene* associated with it (or we might say its inner perimeter of length 2*p*), is expressed by

$$h = p + 3 \qquad (4.19)$$

see also (5), where $n_b'' = 2p$. Now, for a [2p]annulene one finds in analogy with (14)

$$h_{min}^{o} = \begin{cases} \frac{p-1}{2} \; ; & p = 3, 5, 7, 9, \ldots \\ \frac{p}{2} \; ; & p = 6, 8, 10, 12, \ldots \end{cases} \qquad (4.20)$$

The counterpart to eqn. (16) reads

$$h_{max}^{o} = \left[\frac{p^2}{12} + \frac{1}{4} \right] \qquad (4.21)$$

This relation is equivalent to the following equations being analogous with (17).

$$h_{max}^{o}(6j - 3) = \frac{1}{12}(p^2 + 3) = 3j^2 - 3j + 1,$$

$$h_{max}^{o}(6j + \delta) = \frac{1}{12}(p^2 - \delta^2) = 3j^2 + \delta j; \qquad \delta = -2, -1, 0, 1, 2 \qquad (4.22)$$

Here $j = 1, 2, 3, 4, \ldots$, but $j > 1$ for $\delta = -2$. Finally, the numerical values of Table 1 apply to the [2p]annulenes when the relation (19) is taken into account.

PART III

ENUMERATION

Chapter 5

ENUMERATION: GENERAL TREATMENT AND COARSE CLASSIFICATIONS

5.1 INTRODUCTION

By enumeration of coronoid systems the counting of all possible non-isomorphic systems within a class of coronoids is understood. Usually (but not always) the number of hexagons (h) is the leading parameter. The primitive coronoid with $h=8$ (Fig. 2.1) is the smallest coronoid and unique with this number of hexagons. The numbers of non-isomorphic coronoid systems with h = 9, 10 and 11 are 5, 43 and 283, respectively. These numbers increase rapidly, roughly by a factor between 6 and 7 for every h unit, but yet they are known for several additional h values (see below, throughout this chapter). We are not only interested in the grand total of coronoid systems, but also in subdivisions into different classes. In this connection one speaks about classification. This aspect is especially important for large h values.

The enumeration of benzenoids may be traced back to Klarner (1965), who gave the numbers for systems with 1, 2, 3, 4, 5 and 6 hexagons. In his number for $h=6$, viz. 83, if it is computed correctly (which has been questioned in Lunnon 1972), both *hexahelicene* and the quasi-coronoid *coronene* must have been included. The precise interpretation of the numbers under consideration (Klarner 1965; 1967), the so-called Klarner numbers (Knop, Szymanski, Jeričević and Trinajstić 1984) is an interesting question, but outside the scope of the present book.

In some later works coronoid systems have been enumerated together with benzenoids (Knop, Szymanski, Klasinc and Trinajstić 1984; Knop, Szymanski, Jeričević and Trinajstić 1984; Knop, Müller et al. 1985; He WC and He 1985; 1986; 1987; Cioslowski 1987; Balaban, Brunvoll et al. 1987; He WJ et al. 1988; Knop, Müller et al. 1990a; Müller et al. 1990). At this time modern computers were already exploited in this kind of work.

Of special interest in the present context are the specific enumeration data on coronoids, which have appeared (Dias 1982; Knop, Szymanski, Jeričević and Trinajstić 1984; He WC and He 1985; 1986; Knop, Müller et al. 1985; 1986; Cyvin BN, Brunvoll, Cyvin and Gutman 1986; Cyvin SJ, Cyvin and Brunvoll 1987; Brunvoll, Cyvin BN and Cyvin 1987a; Balaban, Brunvoll et al. 1987; He WJ et al. 1988; Cyvin SJ and Brunvoll 1989; Cyvin SJ, Brunvoll and Cyvin 1989c; Knop, Müller et al. 1990a; 1990b; 1990c; Dias 1990; Müller et al. 1990; Cyvin SJ, Brunvoll and Cyvin 1990a). Further references are found in the subsequent chapters, especially for the primitive coronoid systems.

As a general principle, the coronoid (or benzenoid) systems may be generated by adding hexagons to smaller systems. Many options are possible, both with regard to the initial systems and to the addition of hexagons into restricted positions. The choice depends on the problem at hand.

During the computer-generation of coronoid (or benzenoid) systems usually the production of identical or isomorphic systems is inevitable. The computer program must be able to handle the elimination of duplicated systems in one way or another. A procedure to this effect is often the most time-consuming part of the program.

The problems of enumeration of benzenoids and coronoids are to some extent related to the topics of nomenclature and coding of these systems. We shall not be engaged in these topics here, but give some supplementary references, in which coronoid systems are treated or at least mentioned, most of the works being concerned with the nomenclature/coding: Balaban and Harary 1968; Balaban 1969; Elk 1980; Bonchev and Balaban 1981; Elk 1982; Dias 1983; 1984; Elk 1985; Ciosłowski and Turek 1985; Herndon and Bruce 1987; Tošić and Kovačević 1988; Balaban 1988; Kirby 1990.

5.2 GENERAL PRINCIPLES

The computer-aided generations of benzenoid and coronoid systems have been executed in many different ways (Balasubramanian et al. 1980; Knop, Szymanski, Jeričević and Trinajstić 1983; He WC and He 1985; Stojmenović et al. 1986; Brunvoll, Cyvin SJ and Cyvin 1987; Cioslowski 1987). A detailed treatment of these principles, and especially of the different ways for coding the systems, is outside the scope of the present book. We wish only to point out that their name is legion, and it is fascinating to observe how the largely different principles used in the computer programming lead to the same answers. The reader is referred to an extensive survey on the coding by Trinajstić (1990b). Here we show a simple coordinate system applicable to coronoids (and benzenoids; cf. Brunvoll, Cyvin SJ and Cyvin 1987); see Fig. 1. Several of the authors use a definition of the benzenoid by specifying its perimeter.

When it comes to a computer-aided classification of coronoids (or benzenoids) different cases are distinguished (Brunvoll, Cyvin BN and Cyvin 1987b):

(a) specific generation,

(b) recognition,

(c) exclusion.

Suppose that a broad class of coronoids has been generated, e.g., all coronoids with a given h. We speak about *recognition* (b) when a special program or a procedure in the main program detects the members of a subclass within the whole

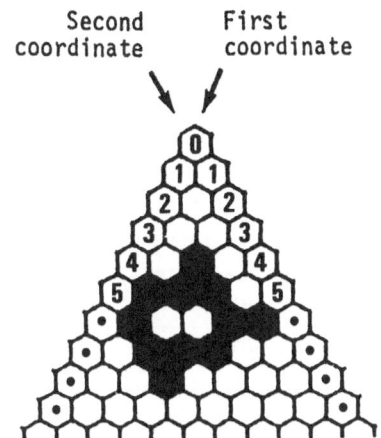

Second coordinate **First coordinate**

Fig. 5.1. A coordinate system for coronoids. In the present example the coronoid is placed into the coordinate system so that it is coded by the following coordinate pairs. (1,5) (2,2) (2,3) (2,4) (3,2) (3,4) (4,1) (4,3) (5,1) (5,2) (5,3)

set. It may be more efficient, however, if it is possible to generate separately a smaller class of coronoids within those with a given h. Then we speak about *specific generation* (a). The principle of *exclusion* (c) is applicable when a class of coronoids in one way or another has been divided into subclasses, but so that one subclass remains. Then the number of coronoids in the last subclass is obviously obtained by subtraction.

We shall add one more to the above principles, viz.

(d) sifting,

and illustrate it by an example. Suppose we have generated all coronoid systems with h hexagons, and we are interested in the systems with $\Delta=1$ and $h+1$ hexagons. Assume that a program is used which generates all the $h+1$ systems, no matter if a system is duplicated. Assume also that the color excess (Δ) can be recognized by a procedure in the program. Then for every generated system it is tested whether $\Delta=1$ or not. If yes, the system is stored; if not, it is discarded. Only the stored systems are compared in order to eliminate isomorphic systems. The great advantage of this so-called *sifting* is, that the comparison of systems is executed within the smaller set of $\Delta=1$ coronoids rather than the whole set of $h+1$ coronoid systems. It should be noted, however, that these systems obtained by sifting can not be used as initial systems in order to produce with certainty all the $\Delta=1$ coronoids with $h+2$ hexagons.

5.3 SURVEY OF POLYHEXES

Here the term *polyhex* is used to denote benzenoids and coronoids together. Benzenoids are polyhexes without holes, while coronoids are polyhexes with holes. It is implied that helicenic systems are not taken into account, an important detail to be emphasized in order to avoid misunderstandings. It means that only geometrically planar polyhexes are considered. It is also important to repeat here that only coronoids with one hole each (single coronoids; cf. Paragraph 2.1.1) are taken into account, which is another limitation of the polyhexes in relation to the usual meaning of this term.

Table 1 shows the available numbers of benzenoids and (single) coronoids along with the sums of these numbers for each h. The documentations in footnotes contain some supplementary references (Cyvin SJ, Cyvin, Brunvoll and Bergan 1987; Müller et al. 1989; Nikolić et al. 1990).

Comment: The number of coronoids with $h = 10$ as found in Knop, Szymanski, Jeričević and Trinajstić 1984 is a misprint, which was reproduced in Knop, Müller et al. 1985. The error was later corrected (Knop, Müller et al. 1986), while the correct number 43 in the meantime also had been derived by others (Brunvoll, Cyvin BN and Cyvin 1987a; Cioslowski 1987).

Table 5.1. Numbers of benzenoids, coronoids and polyhexes (benzenoids + coronoids).

h	Benzenoids	Coronoids	Polyhexes*
1	1[a]		1[b]
2	1[a]		1[b]
3	3[a]		3[b]
4	7[a]		7[b]
5	22[a]		22[b]
6	81[c]		81[b]
7	331[c]		331[b]
8	1435[c]	1[d]	1436[b]
9	6505[c]	5[d]	6510[b]
10	30086[c]	43[e]	30129[f]
11	141229[g]	283[h]	141512[i,j]
12	669584[k]	1954[h]	671538[i]
13	3198256[l]	12363[m,n]	3210619[o]
14	15367577[o-r]	76283[o,s]	15443860[o]
15	74207910[o-q]	453946[o,t]	74661856[o]
16	359863778[o]	2641506[o]	362505284[o]

*Geometrically planar (without helicenic systems).

[a] Klarner DA (1965). Fibonacci Quarterly 3: 9

[b] He WC, He WJ (1985). Theor Chim Acta 68: 301

[c] Knop JV, Szymanski K, Jeričević Ž, Trinajstić N (1983). J Comput Chem 4: 23

[d] Knop JV, Szymanski K, Jeričević Ž, Trinajstić N (1984). Match 16: 119

[e] Knop JV, Müller WR, Szymanski K, Trinajstić N (1986). Match 20: 197

[f] Cioslowski J (1987). J Comput Chem 8: 906

[g] Stojmenović I, Tošić R, Doroslovački R (1986). Proceedings of the Sixth Yugoslav Seminar on Graph Theory, Dubrovnik 1985, Novi Sad: 189

[h] Brunvoll J, Cyvin BN, Cyvin SJ (1987). J Chem Inf Comput Sci 27: 14

[i] He WJ, He WC (1987). Graph Theory and Topology in Chemistry, Elsevier, Amsterdam; Studies in Physical and Theoretical Chemistry 51: 476

[j] Balaban AT, Brunvoll J, Cioslowski J, Cyvin BN, Cyvin SJ, Gutman I, He WC, He WJ, Knop JV, Kovačević M, Müller WR, Szymanski K, Tošić R, Trinajstić N (1987). Z Naturforsch 42a: 863

[k] He WJ, He WC, Wang QX, Brunvoll J, Cyvin SJ (1988). Z Naturforsch 43a: 693

[l] Müller WR, Szymanski K, Knop JV, Nikolić S, Trinajstić N (1989). Croat Chem Acta 62: 481

[m] Cyvin SJ, Cyvin BN, Brunvoll J (1987). Chem Phys Letters 140: 124

[n] Cyvin SJ, Cyvin BN, Brunvoll J, Bergan JL (1987). Coll Sci Papers Fac Sci Kragujevac 8: 137

[o] Knop JV, Müller WR, Szymanski K, Trinajstić N (1990). J Chem Inf Comput Sci 30: 159

[p] Nikolić S, Trinajstić N, Knop JV, Müller WR, Szymanski K (1990). J Math Chem: in press

[q] Knop JV, Müller WR, Szymanski K, Trinajstić N (1990). Reports on Molecular Theory: in press

[r] Müller WR, Szymanski K, Knop JV, Nikolić S, Trinajstić N (1990). J Comput Chem 11: 223

[s] Knop JV, Müller WR, Szymanski K, Trinajstić N (1990). J Mol Struct (Theochem) 205: 361

[t] Brunvoll J, Cyvin BN, Cyvin SJ, Knop JV, Müller WR, Szymanski K, Trinajstić N (1990). J Mol Struct (Theochem) 207: 131

5.4 SOME COARSE CLASSIFICATIONS OF CORONOIDS

5.4.1 *Catacondensed and Pericondensed*

The classes of catacondensed and pericondensed coronoids are defined in Paragraph 3.3.2. The catacondensed systems may be generated specifically by executing successively one-contact additions to the primitive coronoids. The pericondensed systems are obtained by exclusion, i.e. by subtracting from the total. The total numbers are of course obtained (as explained in paragraph 2.1.1) by all types of additions to the primitive coronoids. The five types of addition are defined in Fig. 2.2. Table 2 shows the results of this enumeration.

The pericondensed coronoids contain regular, HED (half essentially disconnected), essentially disconnected and non-Kekuléan systems.

Table 5.2. Numbers of coronoids, divided into catacondensed and pericondensed systems.

h	Catacondensed	Pericondensed	Grand total*
8	1^a		1^c
9	3^a	2^b	5^c
10	15^b	28^b	43^d
11	62^b	221^b	283^b
12	312^e	1642^e	1954^b
13	1435^e	10928^h	$12363^{f,g}$
14	6785^h	69498	$76283^{i,j}$

*See also Table 5.1, especially for $h = 15$ and $h = 16$.

[a]Dias (1982); wrong numbers therein (for $h = 10$, 11 and 12) are omitted.

[b]Brunvoll J, Cyvin BN, Cyvin SJ (1987). J Chem Inf Comput Sci 27: 14

[c]Knop JV, Szymanski K, Jeričević Ž, Trinajstić N (1984). Match 16: 119

[d]Knop JV, Müller WR, Szymanski K, Trinajstić N (1986). Match 20: 197

[e]Balaban AT, Brunvoll J, Cioslowski J, Cyvin BN, Cyvin SJ, Gutman I, He WC, He WJ, Knop JV, Kovačević M, Müller WR, Szymanski K, Tošić R, Trinajstić N (1987). Z Naturforsch 42a: 863

[f]Cyvin SJ, Cyvin BN, Brunvoll J (1987). Chem Phys Letters 140: 124

[g]Cyvin SJ, Cyvin BN, Brunvoll J, Bergan JL (1987). Coll Sci Papers Fac Sci Kragujevac 8: 137

[h]He WJ, He WC, Wang QX, Brunvoll J, Cyvin SJ (1988). Z Naturforsch 43a: 693

[i]Knop JV, Müller WR, Szymanski K, Trinajstić N (1990) J Chem Inf Comput Sci 30: 159

[j]Knop JV, Müller WR, Szymanski K, Trinajstić N (1990) J Mol Struct (Theochem) 205: 361

5.4.2 The rio Classification

The results of enumerations of coronoid systems according to the rio classification (Paragraph 3.3.7) are collected in Table 3.

The non-primitive regular coronoids up to $h = 12$ were obtained by specific generation: the primitive coronoids up to $h = 11$ were subjected to normal additions. When proceeding to $h = 13$, incorporating the primitive coronoids with $h = 12$, the same procedure fails to generate one system, which is shown in Fig. 3.12. If the procedure is to be pursued further one must make proper allowance for possible filling of hexagons into the corona holes of the primitive coronoids. The situation becomes more comprehensible when coronoids with a definite hole are generated specifically at a time. This type of enumeration is treated in some detail in a subsequent chapter.

Non-Kekuléans may easily be recognized among a whole set of coronoids with a given h when $h \leq 14$. Then all of them are obvious non-Kekuléans and therefore reveal themselves by having $\Delta > 0$. Concealed non-Kekulén coronoids, which occur for $h \geq 15$, are recognized by having $\Delta = 0$ and $K=0$.

When the regular (r) and non-Kekuléan (o) systems have been determined and the grand total is known, then the number of irregular (i) coronoids is simply obtained by exclusion.

The regular systems contain the primitive coronoids and non-primitive regular coronoids, both catacondensed and pericondensed. The irregular systems consist of half essentially disconnected and essentially disconnected coronoids, which all are pericondensed. The non-Kekuléan systems are also pericondensed. The regular and irregular systems taken together $(r + i)$ represent all the Kekuléans. The grand total $(r + i + o)$ is consistent with the entries in the last column of Table 2.

Table 5.3. Numbers of coronoids according to the rio classification.[*]

h	r	i	Kekuléan $(r + i)$	o
8	1[a]		1	
9	3[a]		3	2[b]
10	18[a]	6[a]	24	19[c]
11	90[a]	38[a]	128	155[c]
12	526[a]	328[a]	854	1100[c]
13	2810[a]	2240[a]	5050	7313[a]
14	15071	14983	30054	46229

[*] Abbreviations: i irregular; o non-Kekuléan; r regular.
[a] Cyvin SJ, Cyvin BN, Brunvoll J (1987). Chem Phys Letters 140: 124
[b] Cyvin BN, Brunvoll J, Cyvin SJ, Gutman I (1986). Match 21: 301
[c] Brunvoll J, Cyvin BN, Cyvin SJ (1987). J Chem Inf Comput Sci 27: 14

5.4.3 *Classification According to Color Excess*

Benzenoids. For benzenoids it has been demonstrated (Brunvoll, Cyvin BN, Cyvin and Gutman 1988b) that $\Delta_{max} = \left[\dfrac{h}{3}\right]$, where Δ is the color excess (cf. Paragraph 3.2.4). In other words, with increasing h the Δ_{max} value makes a jump for every third h value when $h > 1$: $\Delta=1$ occurs for the first time at $h=3$, $\Delta=2$ at $h=6$, and so on. In general, $\Delta = \Delta_{max}$ occurs for the first time at $h = 3\Delta_{max}$; $\Delta_{max} > 0$. It has also been demonstrated that these extremal obvious non-Kekuléan benzenoids all belong to the class of so-called teepees (or TP benzenoids, where T and P stands for "*Triangulene*" and "*Phenalene*", respectively).

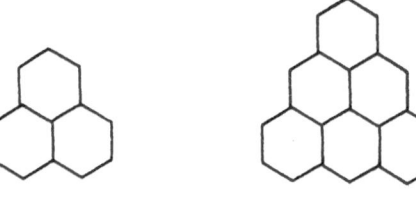

Phenalene   *Triangulene*

A teepee is by definition *phenalene*, *triangulene* or a benzenoid system in which only these two kinds of triangular units are fused (i.e. where two neighbouring units share exactly one edge), and the triangle apex of each unit points the same way (conventionally upwards). Figure 2 shows an illustration.

Color Excess for Coronoids. The properties of Δ_{max} for coronoids are not so regular as for the benzenoids. It is clear that, from an extremal obvious non-Kekuléan system (benzenoid or coronoid) with h' hexagons and color excess $\Delta_{max} = \Delta'$ we may generate a system with $h' + 3$ hexagons and color excess $\Delta' + 1$. This is achieved by fusing a *phenalene* unit to a free edge of the initial system. Such an edge must always exist. Now, in contrast to the situation for benzenoids, it is not certain that the color excess in the generated coronoid system will have the maximum value for the number of hexagons in question. In other words, the Δ_{max} value may jump more frequently than the steps of three h units.

It was actually found that Δ_{max} = 0, 1, 2, 3, 4, 5 and 6 for the first time (lowest h) at h = 8, 9, 10, 11, 13, 16 and 18, respectively; see Table 4.

The selection rules for Δ (Brunvoll, Cyvin BN, Cyvin and Gutman 1988b) are valid for coronoids as well as the benzenoids. Assume that one hexagon is added to a system with parameters h and Δ. Let the corresponding parameters for the new system

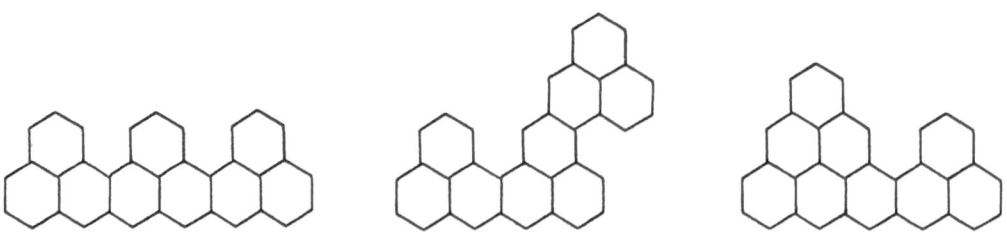

Fig. 5.2. The three existing teepees (teepee benzenoids) with h=9, Δ=3.

Table 5.4. Numbers of coronoids classified according to the color excess.

h	$\Delta = 0$	1	2	3	4	5	6
8	1^a						
9	3^a	2^a					
10	24^a	16^a	3^a				
11	128^a	131^a	23^a	1^a			
12	854^a	906^a	182^a	12^a			
13	5050^b	5913^b	1301^b	98^b	1^b		
14	30054	36555	8875	785	14		
15	†	†	57505	5995	158		
16	†	†	†	42988	1611	4	
17	†	†	†	†	14316	115	
18	†	†	†	†	†	1611	2
19	†	†	†	†	†	†	32

†Unknown

[a]Brunvoll J, Cyvin BN, Cyvin SJ (1987). J Chem Inf Comput Sci 27: 14

[b]He WJ, He WC, Wang QX, Brunvoll J, Cyvin SJ (1988) Z Naturforsch 43a: 693; the number for h=14 (Δ=0) is corrected.

be $h' = h+1$ and Δ'. The selection rules state:

$$\Delta' - \Delta = 0, \pm 1 \tag{5.1}$$

(not −1 if Δ=0). More specifically, if the addition is a one-, three- or five-contact addition (viz. normal addition), then the color excess does not change ($\Delta' = \Delta$). In the case of two- or four-contact additions, the color excess shifts by one unit ($\Delta' = \Delta \pm 1$). These selection rules are useful in enumerations of coronoids with given Δ values, the results of which are shown in Table 4.

For the sake of argument we shall describe how the unique extremal obvious non-Kekuléan coronoid with $h = 13$, Δ=4 (cf. Table 4) can be obtained. All we need in the start are the five coronoid systems with h=9, three with Δ=0 and two with Δ=1. All coronoids with $h = 10$, Δ=1 are generated by non-normal additions to the three systems (h=9, Δ=0) and normal additions to the two systems (h=9, Δ=1). By non-normal additions to the two systems (h=9, Δ=1) we obtain all the (three) systems with $h = 10$, Δ=2. In this run, however, also some systems with $h = 10$, Δ=0 are generated; they may be recognized and discarded. In this way we move diagonally right-downwards in the scheme of Fig. 3. Especially, when the unique system with $h = 11$, Δ=3 is subjected to non-normal additions, the result reveals that no systems with Δ=4 are gene-

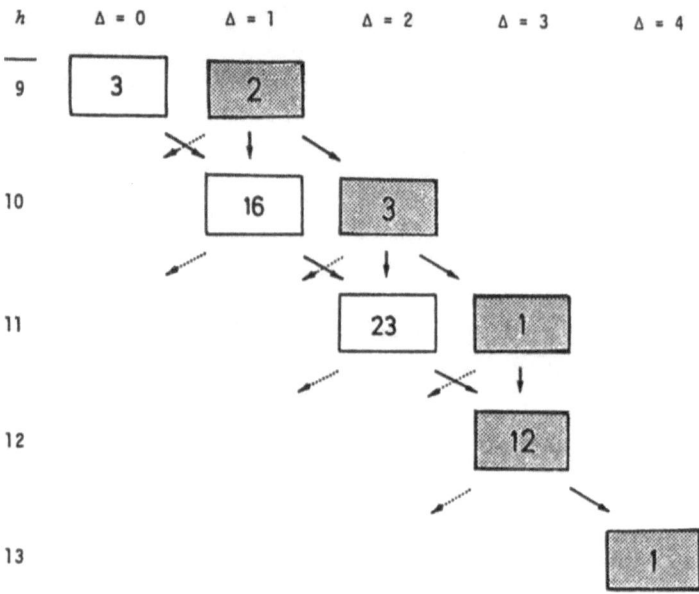

Fig. 5.3. Scheme for the generation of extremal obvious non-Kekuléan coronoid systems, i.e. those with Δ = Δ$_{max}$ (grey rectangles). Vertical arrows: normal (1-, 3- and 5-contact) additions; oblique arrows: non-normal (2- and 4-contact) additions. The scheme ends with the system for h = 13, Δ=4. The entries in rectangles are the numbers of coronoid systems with the appropriate parameters (h, Δ); see also Table 5.4.

rated, only those with Δ=2, which may be discarded.

It has been assumed all the time that the Δ value can be obtained automatically in the computer program. This problem is most easily solved by counting the peaks and valleys (cf. Paragraph 3.2.4).

Forms of Extremal Coronoids. The forms of the extremal obvious non-Kekuléan coronoids are interesting. Those for 9 ≤ h ≤ 14 and the appropriate Δ = Δ$_{max}$ values are depicted in Fig. 4.

Figure 5 continues the illustration of extremal obvious non-Kekuléan coronoids, but only for the (lowest) h values where Δ = Δ$_{max}$ occurs for the first time. This is the case for h = 9, 10, 11, 13, 16, 18, ...

We find that the systems under consideration, where Δ = Δ$_{max}$ is realized for the lowest h value, tend to imitate the teepees inasmuch as the *phenalene* and *triangulene* units are recognized. In addition, however, these systems possess a certain number of corona-condensed (L_2- or A_2-mode) hexagons. We shall refer to such systems as *teepee coronoids* and specifically use the terminology k-teepee, where k indicates the number of corona-condensed hexagons. Some examples are shown in Fig. 6.

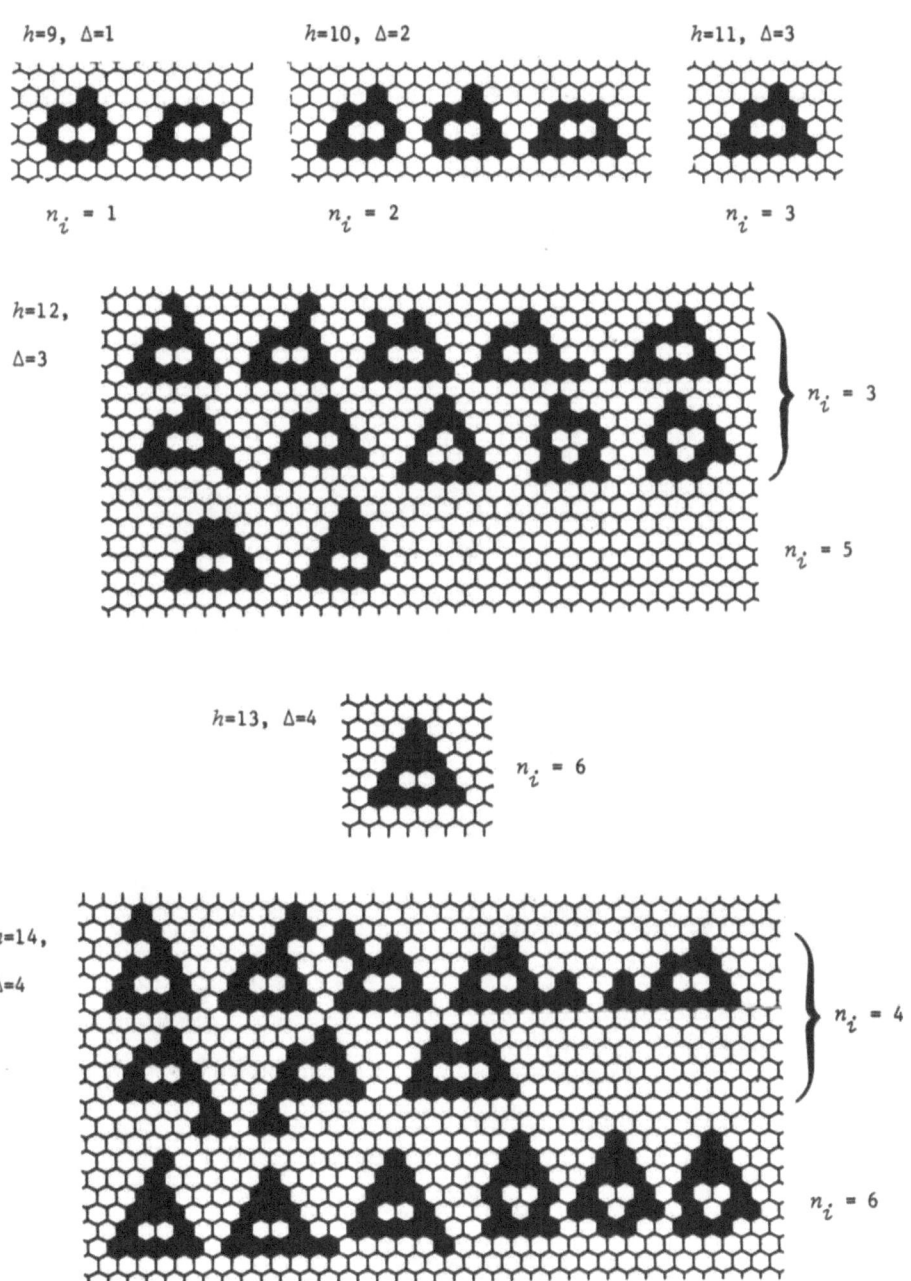

$h=9, \; \Delta=1$ $h=10, \; \Delta=2$ $h=11, \; \Delta=3$

$n_i = 1$ $n_i = 2$ $n_i = 3$

$h=12, \; \Delta=3$

$n_i = 3$

$n_i = 5$

$h=13, \; \Delta=4$ $n_i = 6$

$h=14, \; \Delta=4$

$n_i = 4$

$n_i = 6$

Fig. 5.4. The smallest ($h \leq 14$) extremal obvious non-Kekuléan coronoids, i.e. those with $\Delta = \Delta_{max}$. The numbers of internal vertices (n_i) are given.

In conclusion, it was found by computer generations that the coronoids for $\Delta = \Delta_{max}$ with h = 9, 10, 11, 13, 16 and 18 are 6-, 4-, 2-, 1-, 1- and 0-teepees, respectively. However, k-teepees are also found among the extremal systems for other h values, e.g. the four systems of 3-teepees for h = 12, Δ=3 and the ten systems of 2-teepees for h = 14, Δ=4 (cf. Fig. 4).

Conjectures. The smallest teepee coronoids without a corona-condensed hexagon (viz. 0-teepees) are the two systems with h = 18 (cf. Fig. 5). Hereby it seems reasonable to assume that the situation has become "normalized" in the sense that the Δ_{max} value will increase by one unit for every third h value. Therefore we conjecture for coronoid systems:

$$\Delta_{max} = \left[\frac{h}{3}\right]; \qquad h \geq 18 \qquad\qquad (5.2)$$

when $[x]$ designates the largest integer smaller than or equal to x.

Furthermore, we conjecture that all the extremal obvious non-Kekuléan coronoids with $h = 3\Delta_{max}$ $(\Delta_{max} \geq 6)$ are 0-teepees. Based on this conjecture we have constructed (by trial and error) the coronoids for h = 21, Δ=7 and h = 24, Δ=8 as shown

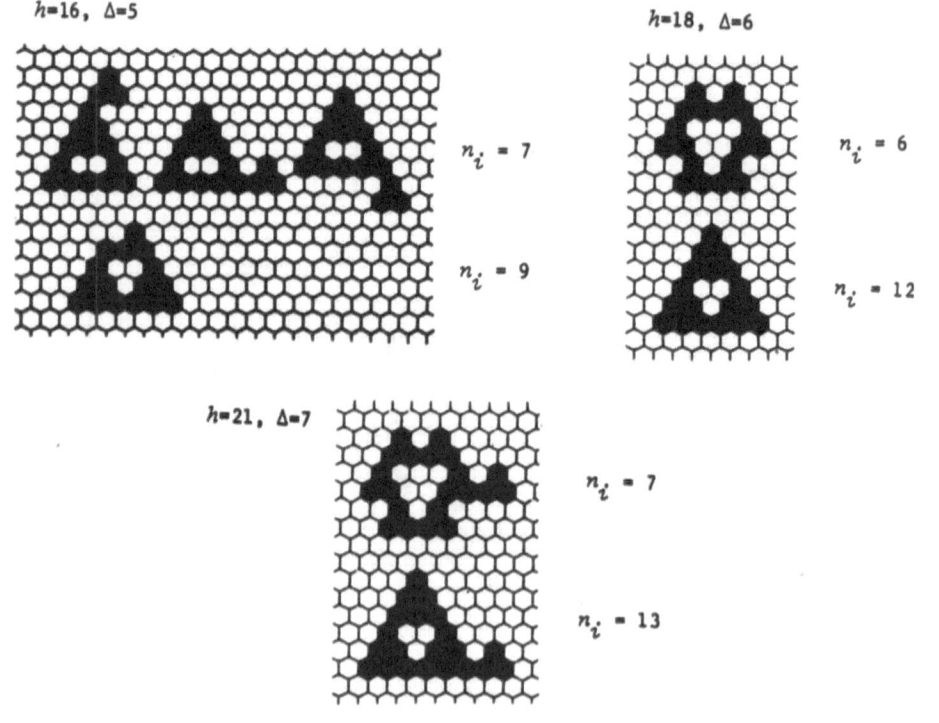

h=16, Δ=5

n_i = 7

n_i = 9

h=18, Δ=6

n_i = 6

n_i = 12

h=21, Δ=7

n_i = 7

n_i = 13

(cont.)

(Fig. 5.5 continued)

$h=24$, $\Delta=8$

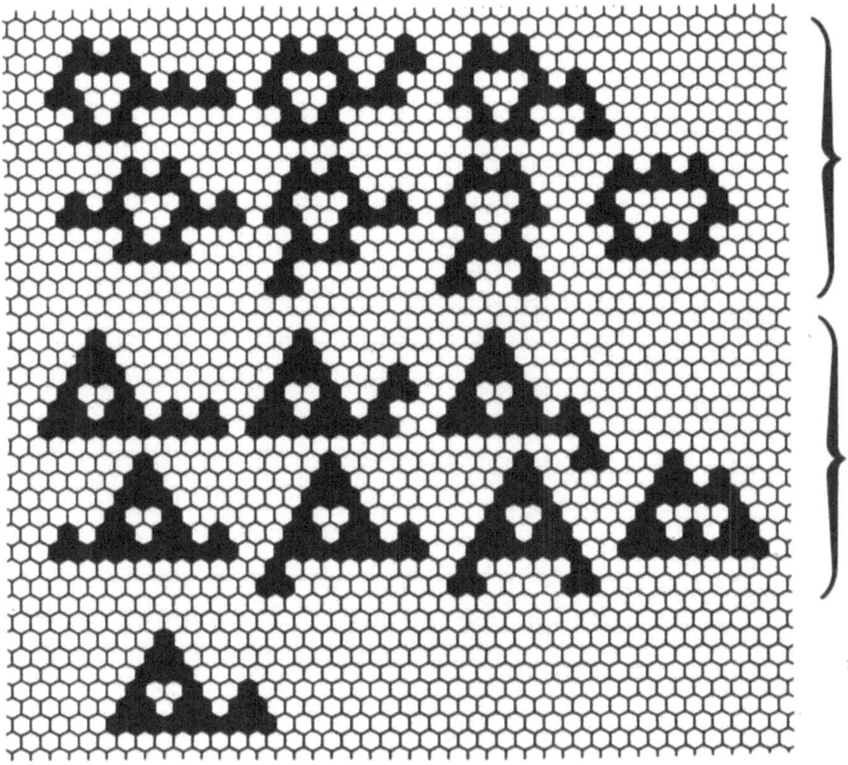

Fig. 5.5. Extremal obvious non-Kekuléan coronoids for h = 16, 18, 21, 24, and the corresponding $\Delta = \Delta_{max}$ values. The numbers of internal vertices (n_i) are given.

in Fig. 5. It is not proved whether they are all the systems of this kind, neither mathematically or by computer analysis. Therefore the corresponding numbers have not been entered into Table 4.

5.4.4 Symmetry

In any set of coronoids (or benzenoids) the symmetry groups of the systems may be determined by recognition (Brunvoll, Cyvin BN and Cyvin 1987b). We shall see that also some details about subdivisions for two of the groups (viz. D_{3h} and C_{2v}) may be determined in this way.

Define the twelve positions obtained by (a) the six rotations as shown in the top row of Fig. 7 and (b) the corresponding six rotations of a mirror image of the coronoid as shown in the bottom row of Fig. 7. Since an unsymmetrical (C_s) coronoid was chosen as example (Fig. 7) all the positions are different. Higher symmetry re-

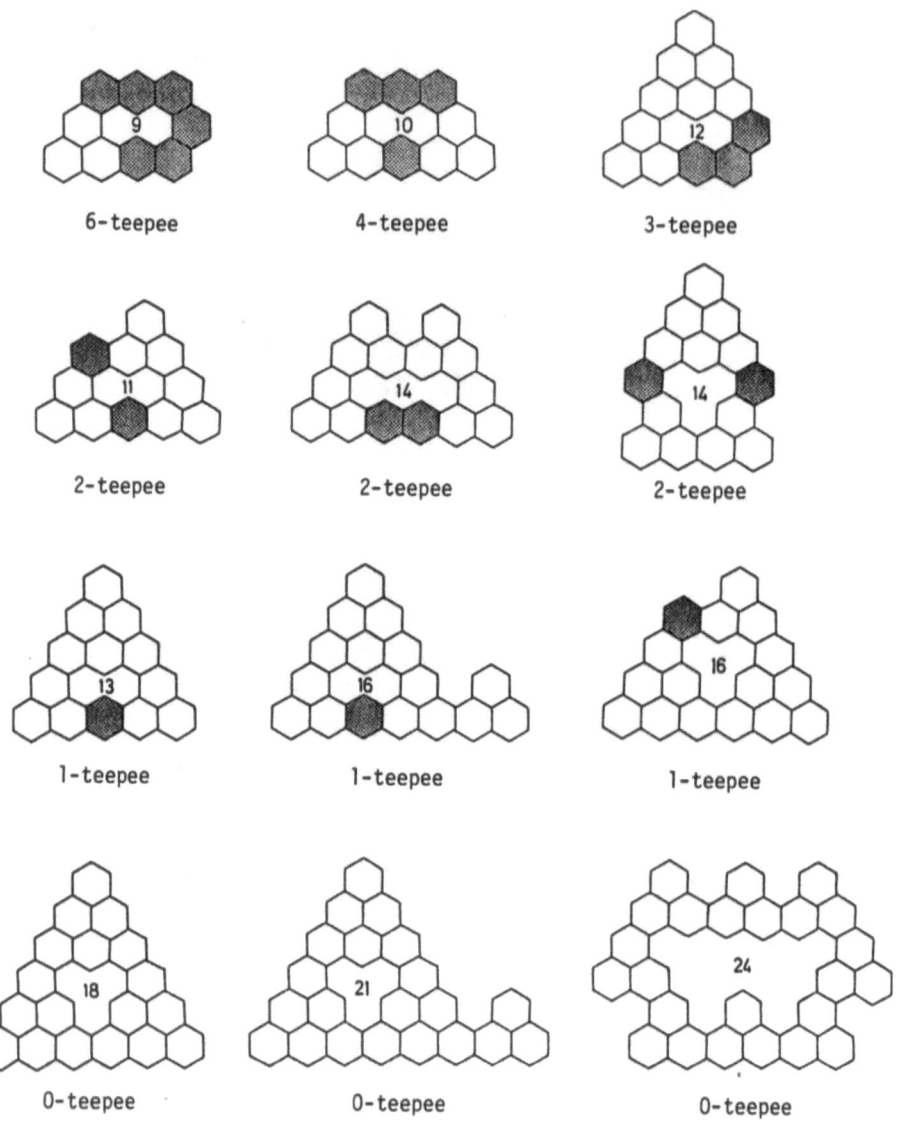

Fig. 5.6. Examples of k-teepees. Corona-condensed hexagons are grey. Numbers of hexagons (h) are inscribed.

veals itself by exhibiting identical pictures in some of the positions:

1. For D_{6h} systems all twelve positions are identical.

    For C_{6h} we find two sets of six identical positions: specifically the Nos. 1, 2, ..., 6 are identical.

2. For D_{3h} systems we find again six identical positions, but now we have Nos. 1, 3, 5, 8, 10, 12 identical in the case of (ia) and 1, 3, 5, 7, 9, 11 identical in the case of (ib) and (ii).

For C_{3h} three positions are identical, e.g. Nos. 1, 3, 5; the same for (i) and (ii).

3. The D_{2h} group is recognized by three sets of four identical positions each. For both kinds (i) and (ii) the positions Nos. 1, 4, 7, 10 are identical. Fig. 7 provides an example if we strip the coronoid for the L_1-mode and the two P_2-mode hexagons.

A C_{2h} system displays six sets of two identical positions each; Nos. 1 and 4 are identical for both kinds (i) and (ii). An example is included in Fig. 7 when we imagine that the L_1-mode hexagon is deleted.

4. For C_{2v} we find also two identical positions, but now Nos. 1 and 10 are identical under the subdivision (a), while Nos. 1 and 7 are identical under (b). The case C_{2v}(a) is illustrated by Fig. 7 if we delete the two P_2-mode hexagons.

5. For C_s, as mentioned above, all positions are different.

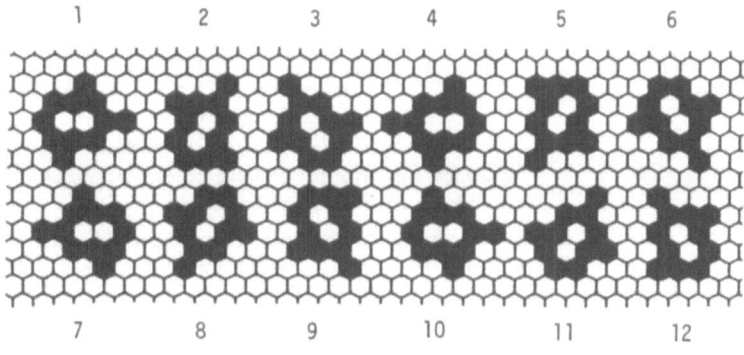

Fig. 5.7. Twelve numbered positions of the coronoid from Fig. 5.1.

Figure 8 shows representatives of coronoids with smallest numbers of hexagons within each symmetry group and their subdivisions.

Table 5 gives the results of enumeration of coronoids with the classification into symmetry groups. Here it is not distinguished between the subdivisions within the groups.

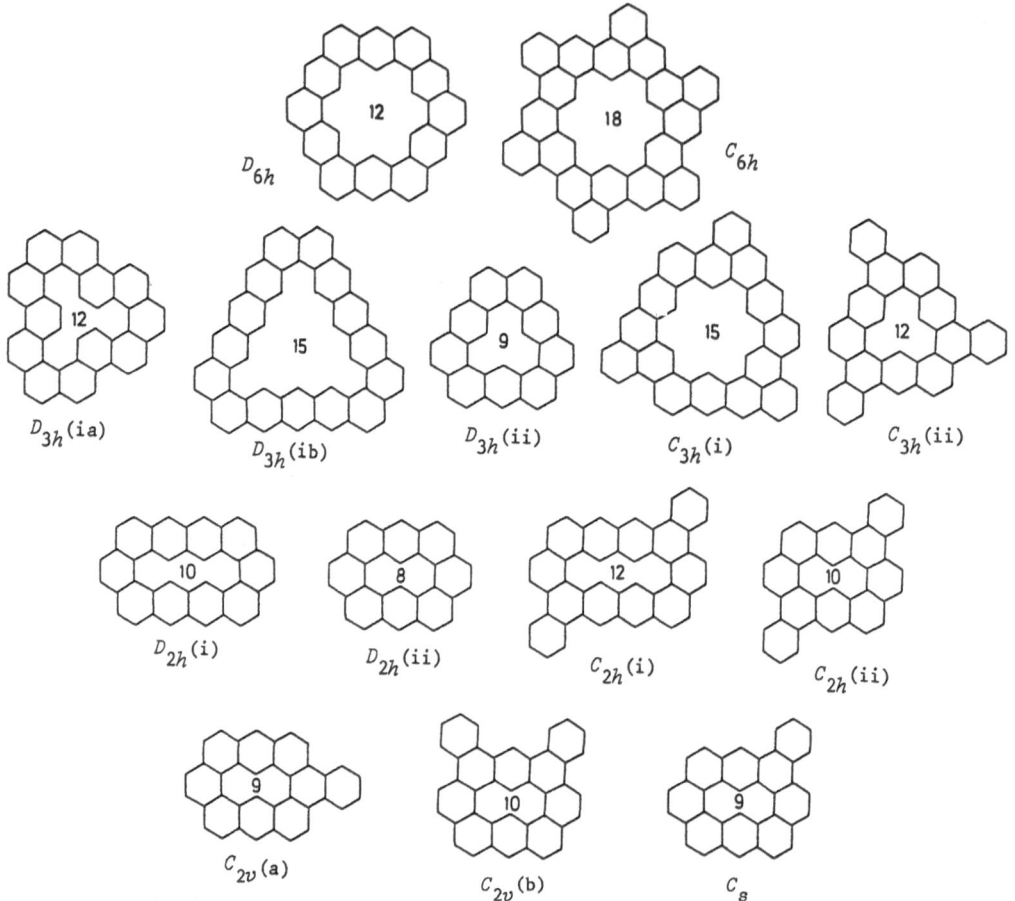

Fig. 5.8. The smallest or one of the smallest representative of coronoids within each symmetry group with their subdivisions. Inscribed numbers are h values.

5.5 CONCLUSION

In the present chapter the main principles of enumeration for coronoids (and benzenoids) were treated. Results of some enumerations and coarse classifications were reported, viz. the numbers of catacondensed/pericondensed coronoids, the *rio* classification, classification according to the color excess, and the distribution into symmetry groups. More detailed enumerations and classifications are found in the subsequent chapters. The next chapter (Chapter 6) gives a detailed account of the primitive coronoids with an appendix on *annulenes*. Chapter 7 is devoted to non-primitive coronoids.

Table 5.5. Numbers of coronoids belonging to the different symmetry groups.

h	D_{6h}	C_{6h}	D_{3h}	C_{3h}	D_{2h}	C_{2h}	C_{2v}	C_s
8					1^a			
9			1^a		0		1^a	3^a
10			0		3^a	3^a	9^a	28^a
11			0		0	0	24^a	259^a
12	1^a		3^a	2^a	10^a	25^a	79^a	1834^a
13	0		0	0	0	0	185^a	12178
14	0		0	0	25	168	579	75511
15	0		9^a	17^a	0	0	†	†
16	0		0	0	†	†	†	†
17	0		0	0	0	0	†	†
18	3^a	1^a	23^a	102^a	†	†	†	†
19	0	0	0	0	0	0	†	†
20	0	0	0	0	†	†	†	†
21	0	0	†	†	0	0	†	†

†Unknown.

a Balaban AT, Brunvoll J, Cioslowski J, Cyvin BN, Cyvin SJ, Gutman I, He WC, He WJ, Knop JV, Kovačević M, Müller WR, Szymanski K, Tošić R, Trinajstić N (1987). Z Naturforsch 42a: 863

Chapter 6

PRIMITIVE CORONOIDS (II) AND ANNULENES: ENUMERATION

6.1. PREVIOUS WORK

As a supplement to the references to previous work listed in Section 5.1 we give here the references to some works which include special treatments of primitive coronoids and their enumeration: Cyvin SJ, Cyvin, Brunvoll and Bergan (1987); Cyvin SJ, Brunvoll and Cyvin (1988; 1989a; 1989d); Brunvoll, Cyvin BN, Cyvin, Gutman, Tošić and Kovačević (1989); Cyvin SJ, Brunvoll, Cyvin, Tošić and Kovačević (1989); Cyvin (1989); Gutman and Cyvin (1989b); Cyvin SJ, Brunvoll, Cyvin, Bergan and Brendsdal (1990).

For treatments of *annulenes* with enumerations, see: Balaban (1971); Cyvin SJ, Brunvoll and Gutman (1990).

6.2 HOLLOW HEXAGONS

6.2.1 *Method of Generation (I)*

Here we present a systematic method of generating by hand (without computer aid) all non-isomorphic hollow hexagons up to a given h. It is referred to Section 4.5 for definitions of the symbols and concepts employed in the following. In particular it should be clear that the hollow hexagons form a subclass of primitive coronoids. The basic principles of the method in question were put forward by Brunvoll, Cyvin BN and Cyvin (1987a).

The distance between two opposite (parallel) segments, say c and f, viz. $\tau_1 = a+b$, is taken as a leading parameter. From a set of *ground forms* with a given τ_1 new systems are derived by increasing the length of c and f in the same pace, i.e. "stretching" the system sideways. If the ground form has h hexagons, the new forms assume $h+2$, $h+4$, $h+6$, ... hexagons. Below we give an example for $\tau_1 = 5$, where new forms with $h = 18$, 20, 22 are derived from a ground form with $h=16$. Dualists are employed. The h values are inscribed into the cycles.

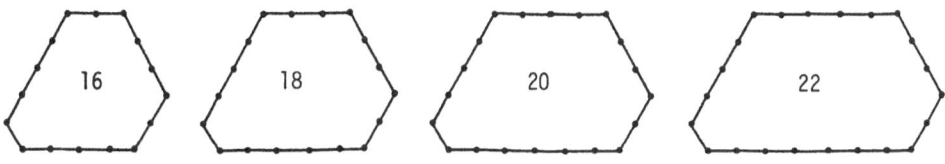

The ground forms for a given τ_1 are determined in the following way.

(1) Determine all the possible combinations of a and b so that $a \geq b$. (2) For every combination, let $c=a$. (3) For every combination, let now d assume all possible values so that $b \leq d \leq a$. (4) Complete the system by e and f, which are uniquely determined.

In the above procedure point (2) is equivalent with $\tau_2 = \tau_1$. Point (3) assures that $\tau_3 \geq \tau_1$. Hence τ_1 becomes the shortest distance, or one of the shortest distances, between the parallel segments.

Figure 1 shows the ground forms of hollow hexagons generated in the described way for τ_1 = 2, 3, 4, 5, 6. Here *coronene* interpreted as a quasi-coronoid with $h=6$ has to be included for the sake of systematization.

For the sake of clarity we show, as examples, the seven and five forms for $h=16$ and 17, respectively (see Fig. 2). For $h=16$ we derive one system each of the basic forms with h even and τ_1 = 2, 3, 4. In addition come the two basic forms for τ_1 = 5. Similarly the systems with $h=17$ are derived from the basic forms with $h=9$ (τ_1 = 3), $h=13$ (τ_1 = 4) and $h=15$ (τ_1 = 5), in addition to the basic form itself for $h=17$. In this way all hollow hexagons for $h \leq 20$ are obtainable from Fig. 1. An extension of the figure to $\tau_1 = \tau_1^*$ would provide all the systems of hollow hexagons up to $3\tau_1^* + 2$.

6.2.2 *Symmetry Considerations*

As mentioned in Section 4.5 a hollow hexagon belongs to one of the symmetry groups D_{6h}, D_{3h}, D_{2h}, C_{2h}, C_{2v} or C_s. When they are arranged as in Fig. 1, exactly all which belong to the regular hexagonal (D_{6h}) and trigonal (D_{3h}) symmetries are found in the first column. These trigonal systems all have a vertical two-fold symmetry axis; they belong to either (ia) or (ii). At the end (right-hand side) of each row (Fig. 1), if the system is not hexagonal, it is dihedral (D_{2h}). Both kinds occur, viz. (i) and (ii). The rest of the ground forms of hollow hexagons are mirror-symmetrical (C_{2v}), to be classified as (b).

When a regular hexagonal hollow hexagon is stretched (as described in the generation procedure above), it becomes D_{2h}, either (i) or (ii). A stretched regular trigonal system is C_{2v}(b). Each of the dihedral systems are stretched into centrosymmetry, viz. C_{2h}, either (i) or (ii). Finally a mirror-symmetrical ground form becomes stretched into an unsymmetrical (C_s) system.

For explanations of the symmetry classifications, see Paragraph 3.3.8.

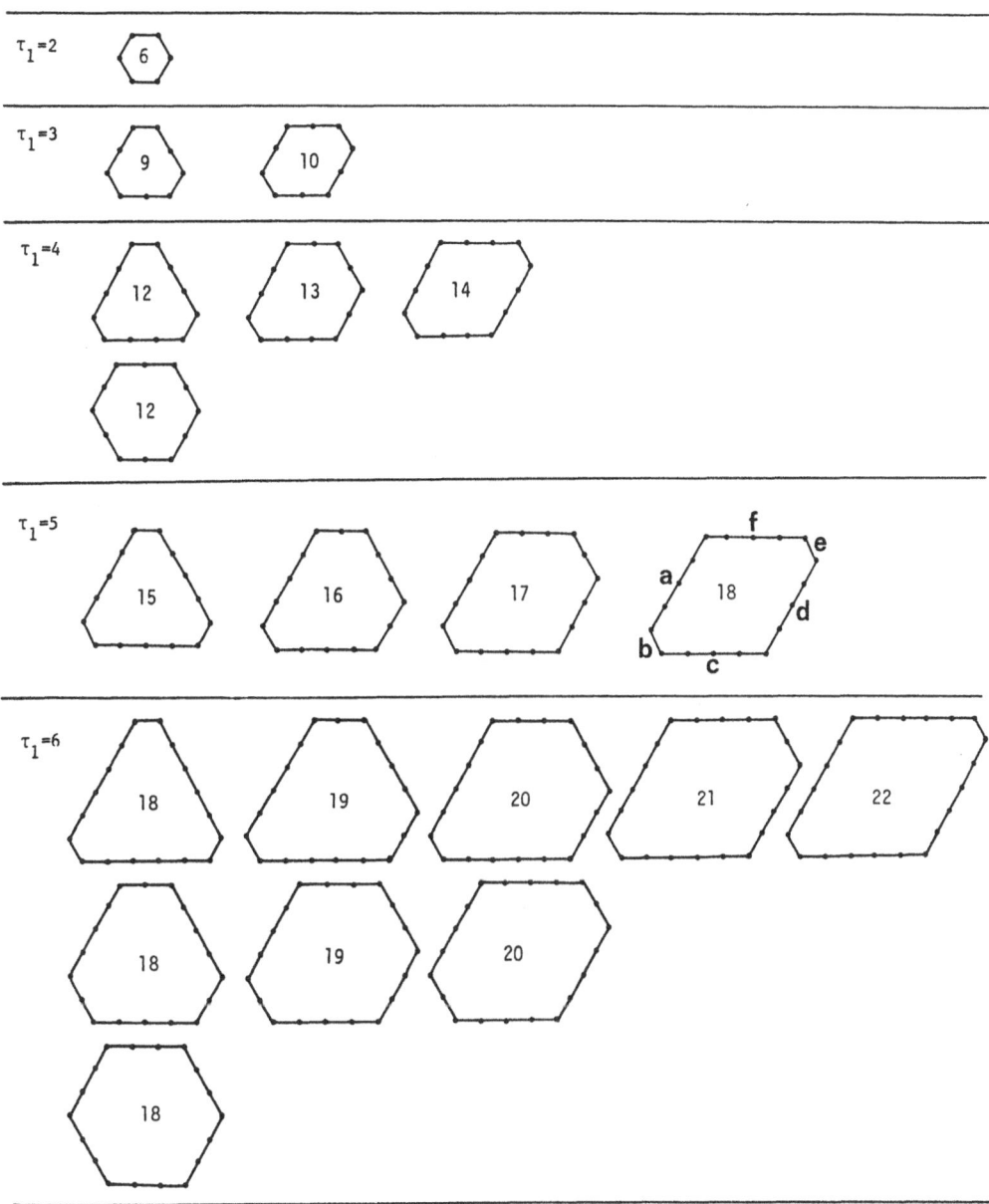

Fig. 6.1. Ground forms of hollow hexagons for $2 \leq \tau_1 \leq 6$. Dualists are employed. The inscribed figures are h values.

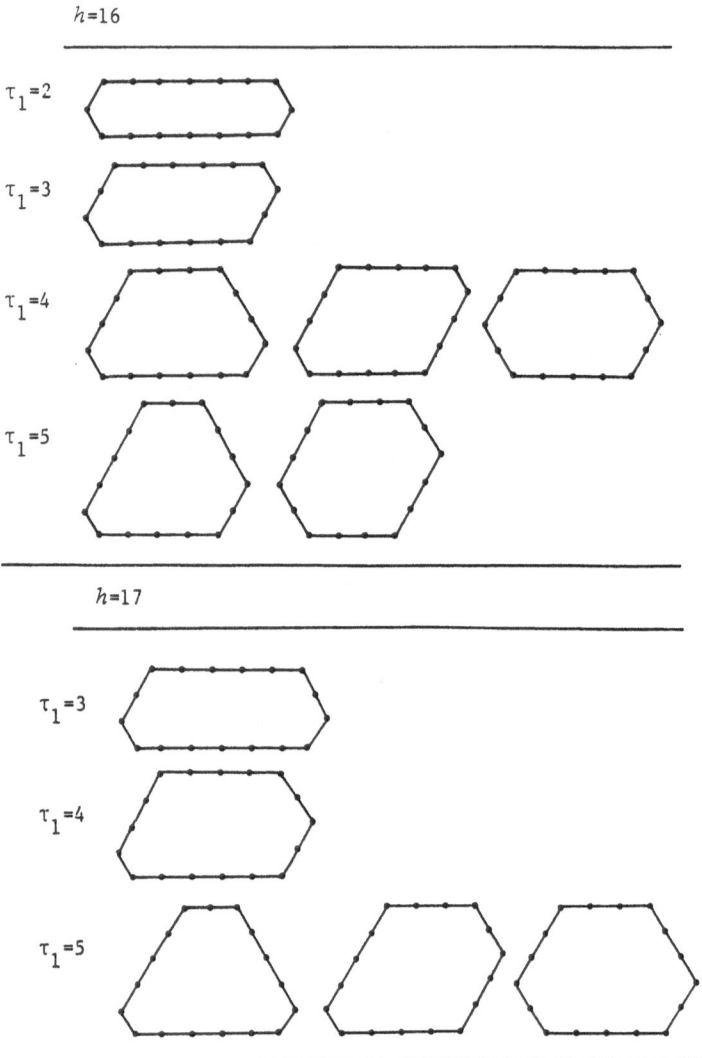

Fig. 6.2. All hollow hexagons (as dualists) for $h = 16$ and $h = 17$.

6.2.3 *Enumeration Results*

Based on the procedure described above the following explicit formulas were deduced (Cyvin SJ, Brunvoll and Cyvin 1989d) for the number of non-isomorphic hollow hexagons as a function of h, say $N_{Hh}(h)$.

$$N_{Hh}(6j) = \frac{1}{8}\left[(j+1)(2j^2 + j + 1) - \frac{1 + (-1)^j}{2}\right],$$

$$N_{Hh}(6j+2) = \frac{1}{8}\left[(j+1)(2j^2 + 3j - 1) + \frac{1 + (-1)^j}{2}\right],$$

$$N_{Hh}(6j+4) = \frac{1}{8}\left[(j+1)(2j^2 + 5j + 1) - \frac{1 + (-1)^j}{2}\right]; \quad j = 1, 2, 3, \ldots \quad (6.1)$$

These equations cover all cases for even h, including the quasi-coronoid *coronene*, for which $j=1$ gives $N_{Hh}(6) = 1$. The numbers for odd h values are obtainable from (1) together with

$$N_{Hh}(h+3) = N_{Hh}(h); \quad h = 6, 8, 10, \ldots \quad (6.2)$$

Table 1 shows the numerical values of $N_{Hh}(h)$ for h up to 60.

The symmetry properties (see above) were utilized in an analysis of the symmetry distribution of the hollow hexagons (Cyvin SJ, Brunvoll, Cyvin, Bergan and Brendsdal 1990). The results are presented in Table 2.

Table 6.1. Numbers of hollow hexagons, $N_{Hh}(h)$, for $h \leq 60$.

h		$N_{Hh}(h)$	h		$N_{Hh}(h)$
(6)*	9	1	34	37	57
8	11	1	36	39	69
10	13	2	38	41	78
12	15	4	40	43	90
14	17	5	42	45	106
16	19	7	44	47	118
18	21	11	46	49	134
20	23	13	48	51	154
22	25	17	50	53	170
24	27	23	52	55	190
26	29	27	54	57	215
28	31	33	56	59	235
30	33	42	58		260
32	35	48	60		290

*Quasi-coronoid (*coronene*).

Table 6.2. Numbers of hollow hexagons, classified according to symmetry.* (Parenthesized values do not pertain to genuine coronoids, but conform with the formula apparatus.)

h	D_{6h}	D_{3h}	D_{2h}	C_{2h}	C_{2v}	C_s	Total
(6)	(1)	(0)	(0)	(0)	(0)	(0)	(1)
(7)	(0)	(0)	(0)	(0)	(0)	(0)	(0)
8	0	0	1	0	0	0	1
9	0	1	0	0	0	0	1
10	0	0	2	0	0	0	2
11	0	0	0	0	1	0	1
12	1	1	1	1	0	0	4
13	0	0	0	0	2	0	2
14	0	0	3	1	1	0	5
15	0	2	0	0	1	1	4
16	0	0	3	2	2	0	7
17	0	0	0	0	4	1	5
18	1	2	3	3	1	1	11
19	0	0	0	0	5	2	7
20	0	0	4	4	4	1	13
21	0	3	0	0	4	4	11
22	0	0	5	5	5	2	17
23	0	0	0	0	8	5	13
24	1	3	4	7	4	4	23
25	0	0	0	0	10	7	17
26	0	0	6	8	8	5	27
27	0	4	0	0	8	11	23
28	0	0	6	10	10	7	33
29	0	0	0	0	14	13	27
30	1	4	6	12	8	11	42
$6j$	1	$j-1$	$\frac{1}{2}[3(j-1)-\varepsilon]$	$\frac{1}{4}[3(j-1)^2+\varepsilon]$	$\frac{1}{4}(j-1)(3j-7)+\varepsilon$	$\frac{1}{8}(j-1)(2j^2-7j+7)-\varepsilon$	$\frac{1}{8}(j+1)(2j^2+j+1)-\varepsilon$
$6j+1$	0	0	0	0	$\frac{1}{4}(j-1)(3j+1)+\varepsilon$	$\frac{1}{8}(j-1)(2j^2-3j-1)-\varepsilon$	$\frac{1}{8}(j+1)(j-1)(2j+1)+\varepsilon$
$6j+2$	0	0	$\frac{1}{2}(3j-1+\varepsilon)$	$\frac{1}{4}(j-1)(3j-1)-\varepsilon$	$\frac{1}{4}(j-1)(3j-1)+\varepsilon$	$\frac{1}{8}(j-1)(2j^2-5j+1)+\varepsilon$	$\frac{1}{8}(j+1)(2j^2+3j-1)+\varepsilon$
$6j+3$	0	j	0	0	$\frac{1}{4}(j-1)(3j-1)-\varepsilon$	$\frac{1}{8}(j-1)(2j^2-j+1)+\varepsilon$	$\frac{1}{8}(j+1)(2j^2+j+1)-\varepsilon$
$6j+4$	0	0	$\frac{1}{2}(3j+1-\varepsilon)$	$\frac{1}{4}(j-1)(3j+1)+\varepsilon$	$\frac{1}{4}(j-1)(3j+1)+\varepsilon$	$\frac{1}{8}(j-1)(2j^2+5j+1)-\varepsilon$	$\frac{1}{8}(j+1)(2j^2+5j+1)-\varepsilon$
$6j+5$	0	0	0	0	$\frac{1}{4}(j+1)(3j-1)+\varepsilon$	$\frac{1}{8}(j-1)(j+1)(2j-1)-\varepsilon$	$\frac{1}{8}(j+1)(2j^2+3j-1)+\varepsilon$

*In the analytical expressions, $j = 1,2,3,\ldots\ldots$; $\varepsilon = \frac{1}{2}[1+(-1)^j]$.

6.2.4 *Method of Generation (II)*

The hollow hexagons were also generated and enumerated by a completely different method, amenable for computer programming. This method is purely combinatorial: Three integers (a, b, c) are generated in all possible ways which give $d > 0$ according to eqn. (4.12), where h is a prescribed number. As before e and f are determined by (4.11). Isomorphic systems were eliminated by realizing that reversion and cyclic permutations of the parameters are equivalent. Simple criteria were used to determine the symmetries of the generated systems. As expected, the numerical results of Tables 1 and 2 were all confirmed.

6.2.5 *Forms*

·Figure 3 shows the generated forms of hollow hexagons up to $h=20$. They are supplied with K numbers and actually ordered according to increasing K.

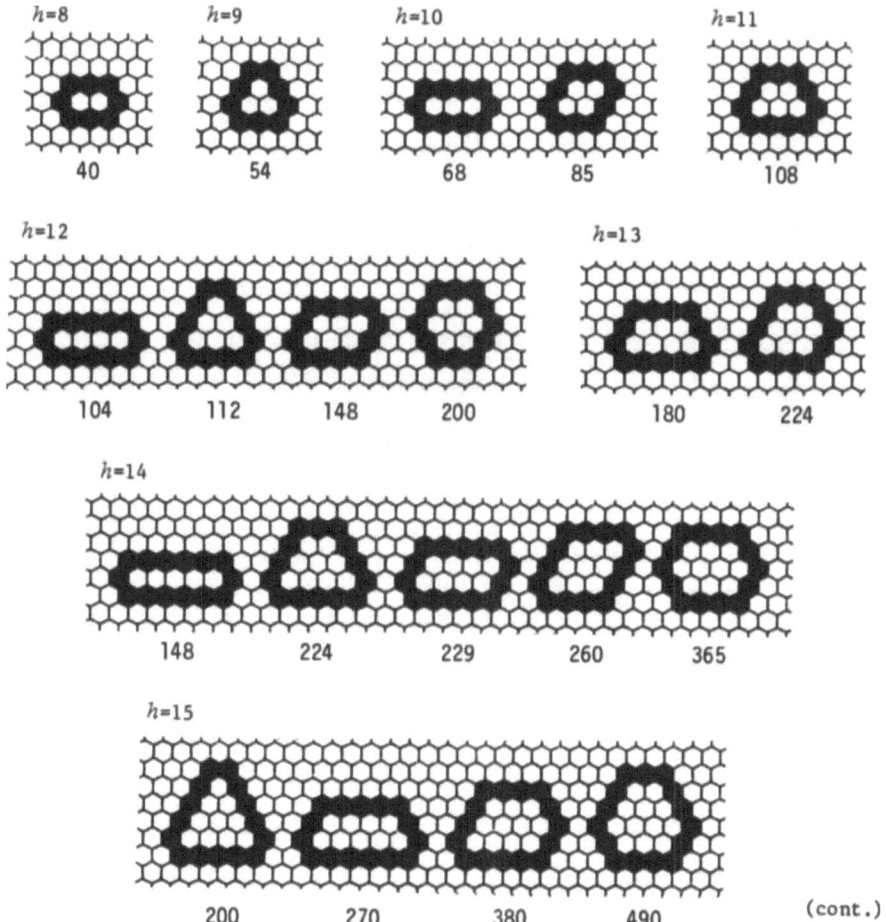

(cont.)

(Fig. 6.3 continued)

$h=16$

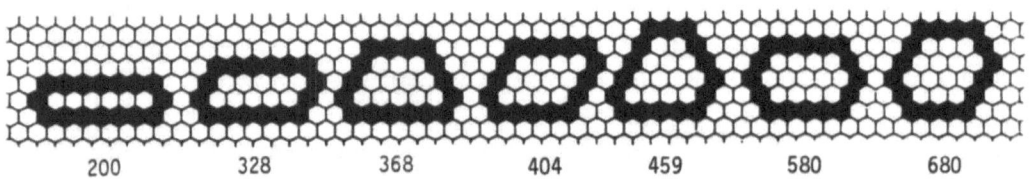

200   328   368   404   459   580   680

$h=17$

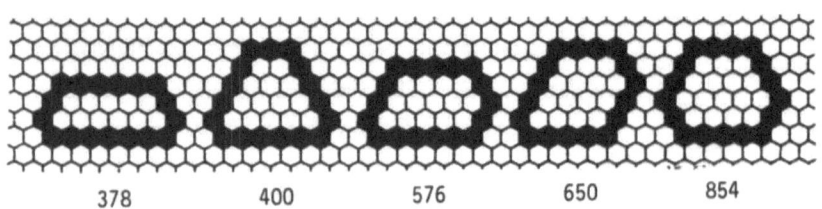

378   400   576   650   854

$h=18$

260   324   445   544   580   629

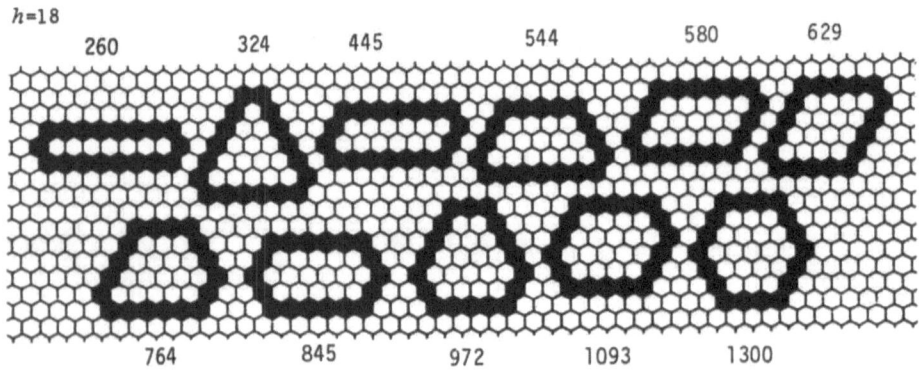

764   845   972   1093   1300

$h=19$

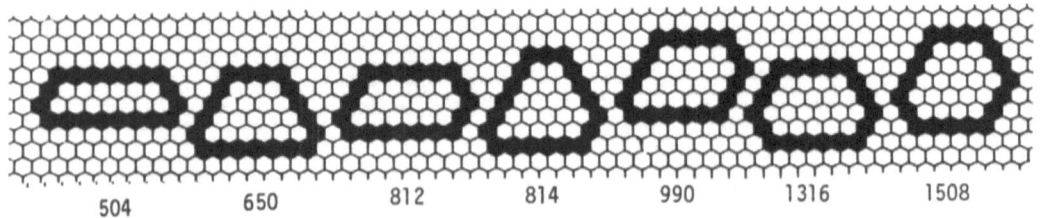

504   650   812   814   990   1316   1508

(cont.)

(Fig. 6.3 continued)

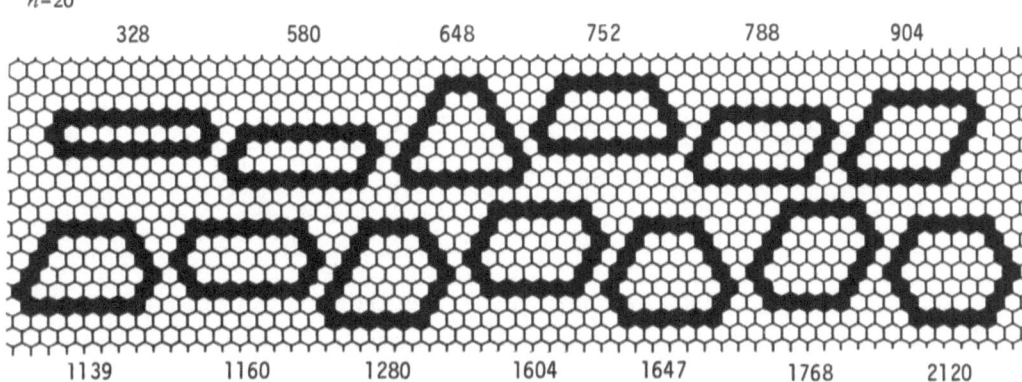

$h=20$

Fig. 6.3. All (non-isomorphic) hollow hexagons for $8 \le h \le 20$. K numbers are given.

6.3 PRIMITIVE CORONOIDS

6.3.1 *Method of Generation (I)*

Brunvoll, Cyvin BN and Cyvin (1987a) suggested a method of constructing all primitive coronoids with a given h by modifications of the hollow hexagons with the same h. This modification implies the creation of intruding corners (for definitions here and throughout this section, see Section 4.2).

The most typical modification just consists of moving one hexagon in order to convert a protruding corner into an intruding one. The modifications are supposed to be executed in several steps as far as possible. Example:

This method must also allow for a moving of more than one hexagon at a time, for example:

In this last example the numbers of protruding and intruding corners are actually the same before and after the modification, but intruding corners have of course been created in relation to the original hollow hexagon.

In Table 3 the twenty first hollow hexagons (for $h \leq 15$) are depicted as dualists. In each case the number of primitive coronoids derived from the hollow hexagon (modified systems and the hollow hexagon itself) is given.

The above outlined method by modifications of hollow hexagons has proved to be useful in practice for hand-generation of primitive coronoids with the aid of a portion of intuition. However, the procedure is not precisely defined in general terms so as to lend itself to computer programming. In the subsequent paragraph we describe another method, which has been adapted to a computer program.

6.3.2 *Method of Generation (II)*

Brunvoll, Cyvin BN, Cyvin, Gutman, Tošić and Kovačević (1989) devised a procedure for generating the primitive coronoids from the corona hole.

A primitive coronoid is uniquely determined by its hole. The systematic generation of the holes should, for $h < 22$, take into account all benzenoids with at least two hexagons and without any coves and fjords. More about the restrictions on corona holes for primitive coronoids is found in Paragraph 4.2.2.

The corona holes are generated by additions of hexagons in the usual way for benzenoids, as may be symbolized by $h^o \rightarrow h^o + 1$. Only three of the five possible additions (cf. Fig. 2.2) may come into operation here. It should be observed that the number of hexagons of the primitive coronoid, viz. h, does not increase in the same pace as the number of hexagons of the corona hole (h^o). The situation is summarized in the following.

(i) One-contact addition; $h \rightarrow h+2$,

(ii) two-contact addition; $h \rightarrow h+1$,

(iii) three-contact addition; $h \rightarrow h+0$.

All the three possibilities (i), (ii) and (iii) are necessary, as is demonstrated in Fig. 4, which shows the possible ways of generating the holes of primitive coronoids up to $h=12$. In particular it is seen that (iii) becomes necessary for the

Table 6.3. Hollow hexagons for $h \leq 15$ and the numbers of primitive coronoids.

h	No.	Form	Number of systems derived from the hollow hexagon	Total number of primitive coronoids
8	1		1	1
9	2		1	1
10	3		1	3
	4		2	
11	5		2	2
12	6		1	11
	7		1	
	8		4	
	9		5	
13	10		4	12
	11		8	
14	12		1	40
	13		6	
	14		3	
	15		21	
	16		9	
15	17		6	68
	18		42	
	19		1	
	20		19	

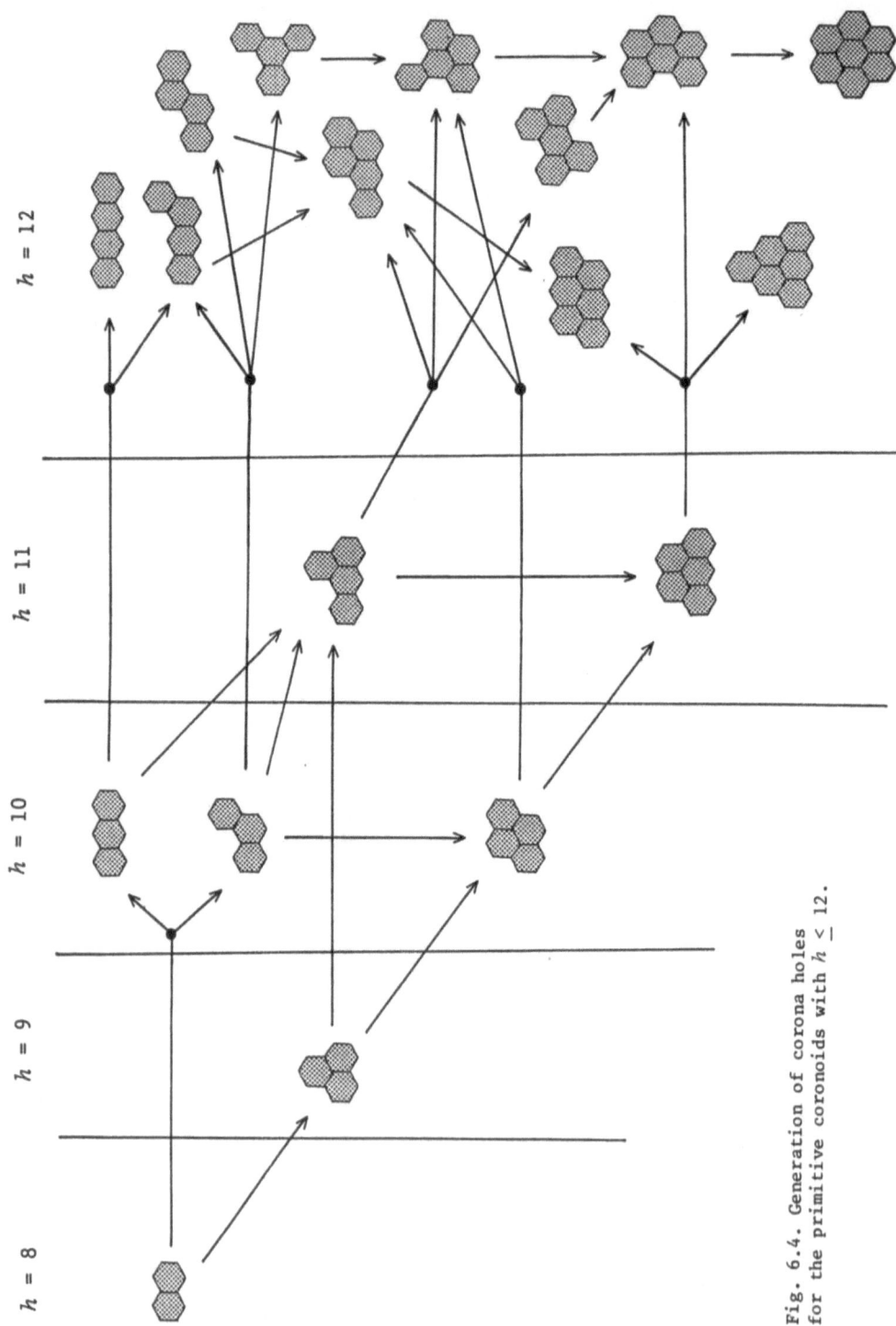

Fig. 6.4. Generation of corona holes
for the primitive coronoids with $h \leq 12$.

$h = 12$

$h = 11$

$h = 10$

$h = 9$

$h = 8$

Table 6.4. Numbers of primitive coronoids with different sizes of the corona hole.*

h^o = number of hexagons in the corona hole

h	2	3	4	5	6	7	8	9	10	11	12	13	14	15	16	17	18	19	20	21	22	23	24	25	26	27	Total
8	1																										1
9		1																									1
10		2	1																								3
11			1	1																							2
12			4	3	3	1																					11
13				4	4	3	1																				12
14				8	9	9	9	4	1																		40
15					12	14	16	14	9	2	1																68
16					16	26	35	36	36	25	13	4	1														192
17						37	49	65	58	68	53	34	16	4	1												395
18						40	86	122	144	158	161	134	106	67	30	9	3	1									1061
19							111	187	261	291	330	331	298	249	188	111	52	18	4	1							2432
20							92	282	475	589	677	740	772	710	640	526	386	236	125	47	16	4	1				6318
21								340	666	1081	1297	1479	1613	1693	1618	1504	1317	1060	780	512	274	124	43	13	2	1	15417

*Brunvoll J, Cyvin BN, Cyvin SJ, Gutman I, Tošić R, Kovačević M (1989). J Mol Struct (Theochem) 184: 165

first time when *coronene* is to be generated; this is the corona hole of *kekulene*.

It has already been demonstrated (Fig. 4) that a corona hole may be generated in several ways. The program must take care of eliminating all systems isomorphic with those already generated, no matter in which way this duplication has arised. The program must also avoid creation of coves and fjords of the corona holes (but not the coronoids). It must also take proper care of the book-keeping for h values.

Table 4 shows the computational results up to $h=21$ according to the procedure described above. The sizes of the corona hole are accounted for. It is referred to Section 4.6 for an account of the h_{min}^{o} and h_{max}^{o} values as functions of h.

6.3.3 *Extended Computations*

Another computer program for the enumeration of primitive coronoids was designed (Brunvoll, Cyvin BN, Cyvin, Gutman, Tošić and Kovačević 1989). It is based on a more direct method, which allows the generation for a given h value without invoking the smaller systems. This analysis did not give any information of the size of the corona hole in supplement to Table 4, but the total numbers of primitive coronoids were supplemented by the data up to $h \leq 27$.

Table 5 summarizes the numbers of primitive coronoids for $h \leq 27$. The symmetry distribution was obtained by both approaches, the extended computations as well as the method of the preceding paragraph. The pertinent data are included in Table 5.

6.3.4 *Forms*

In Fig. 5 we show the forms of the primitive coronoids with $h \leq 15$, ordered again according to increasing K numbers. These systems correspond exactly to those which first were derived by hand according to the method described in Section 6.3.1 (see especially Table 3).

Table 6.5. Numbers of primitive coronoids classified according to symmetry.

h	D_{6h}	C_{6h}	D_{3h}	C_{3h}	D_{2h}	C_{2h}	C_{2v}	C_s	Total
8	0	0	0	0	1[a]	0	0	0	1[a]
9	0	0	1[b]	0	0	0	0	0	1[a]
10	0	0	0	0	2[c]	0	1[c]	0	3[a]
11	0	0	0	0	0	0	1[c]	1[c]	2[a]
12	1[d,e]	0	2[b]	0	2[c]	2[c]	2[c]	2[c]	11[a]
13	0	0	0	0	0	0	4[c]	8[c]	12[a]
14	0	0	0	0	6[c]	5[c]	12[c]	17[c]	40[a]
15	0	0	2[b]	2[b]	0	0	6[c]	58[c]	68[f]
16	0	0	0	0	7[c]	18[c]	38[c]	129[c]	192[f]
17	0	0	0	0	0	0	23[c]	372[c]	395[f]
18	2[d,e]	0	5[b]	4[b]	15[c]	48[c]	92[c]	895[c]	1061[f]
19	0	0	0	0	0	0	55[c]	2377[c]	2432[f]
20	0	0	0	0	20[c]	137[c]	272[c]	5889[c]	6318[f]
21	0	0	5[b]	15[b]	0	0	142[c]	15255[c]	15417[c,g]
22	0	0	0	0	46[c]	363[c]	705[c]	38546[c]	39660[c]
23	0	0	0	0	0	0	367[c]	99183[c]	99550[c]
24	2[d,e]	1[d,e]	17[b]	34[b]	50[c]	992[c]	1872[c]	253420[c]	256388[c]
25	0	0	0	0	0	0	973[c]	653030[c]	654003[c]
26	0	0	0	0	125[h]	2629[h]	4930[h]	1682289[h]	1689973[h]
27	0	0	13[b]	104[b]	0	0	2470[h]	4352870[h]	4355457[h]

[a] Brunvoll J, Cyvin BN, Cyvin SJ (1987). J Chem Inf Comput Sci 27: 14

[b] Cyvin SJ, Brunvoll J, Cyvin BN (1988). Acta Chem Scand A42: 434

[c] Brunvoll J, Cyvin BN, Cyvin SJ, Gutman I, Tošić R, Kovačević M (1989). J Mol Struct (Theochem) 184: 165

[d] Cyvin SJ, Cyvin BN, Brunvoll J, Bergan JL (1987). Coll Sci Papers Fac Sci Kragujevac 8: 137

[e] Cyvin SJ (1989). Monatsh Chem 120: 243

[f] Balaban AT, Brunvoll J, Cioslowski J, Cyvin BN, Cyvin SJ, Gutman I, He WC, He WJ, Knop JV, Kovačević M, Müller WR, Szymanski K, Tošić R, Trinajstić N (1987). Z Naturforsch 42a: 863

[g] He WJ, He WC, Wang QX, Brunvoll J, Cyvin SJ (1988). Z Naturforsch 43a: 693

[h] Computed by a program from R. Tošić and M. Kovačević.

$h=8$ $\qquad$ $h=9$ $\qquad$ $h=10$

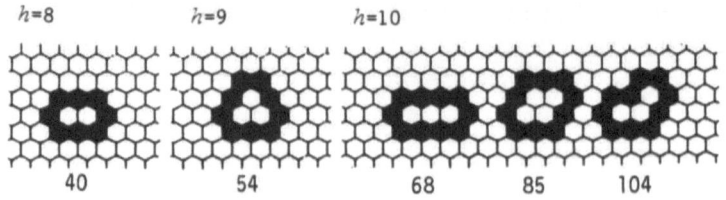

40 $\qquad$ 54 $\qquad$ 68 $\qquad$ 85 $\qquad$ 104

$h=11$

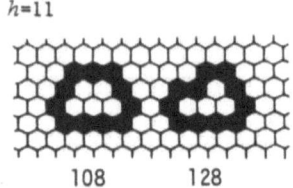

108 $\qquad$ 128

$h=12$

104 $\qquad$ 112 $\qquad$ 148 $\qquad$ 197 $\qquad$ 200

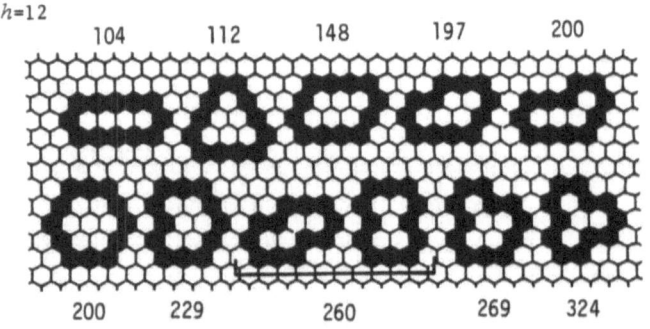

200 $\qquad$ 229 $\qquad$ 260 $\qquad$ 269 $\qquad$ 324

$h=13$

180 $\qquad$ 224 $\qquad$ 230 $\qquad$ 236 $\qquad$ 252 $\qquad$ 288

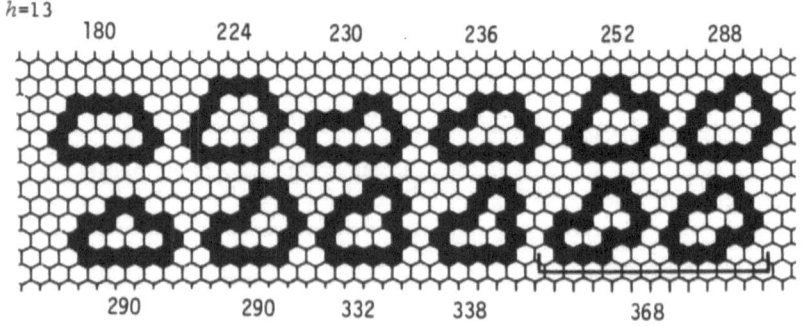

290 $\qquad$ 290 $\qquad$ 332 $\qquad$ 338 $\qquad$ 368

(cont.)

(Fig. 6.5 continued)

$h=14$

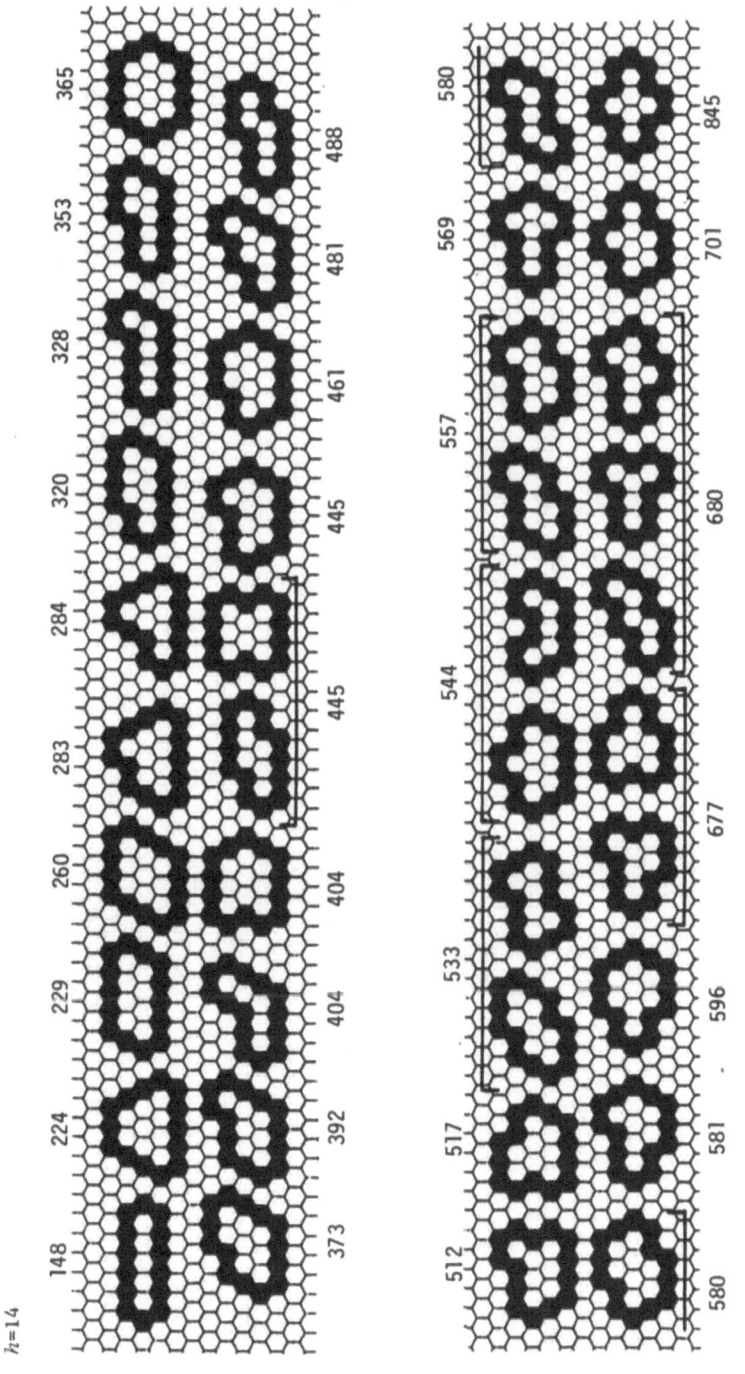

(cont.)

(Fig. 6.5 continued)

h=15

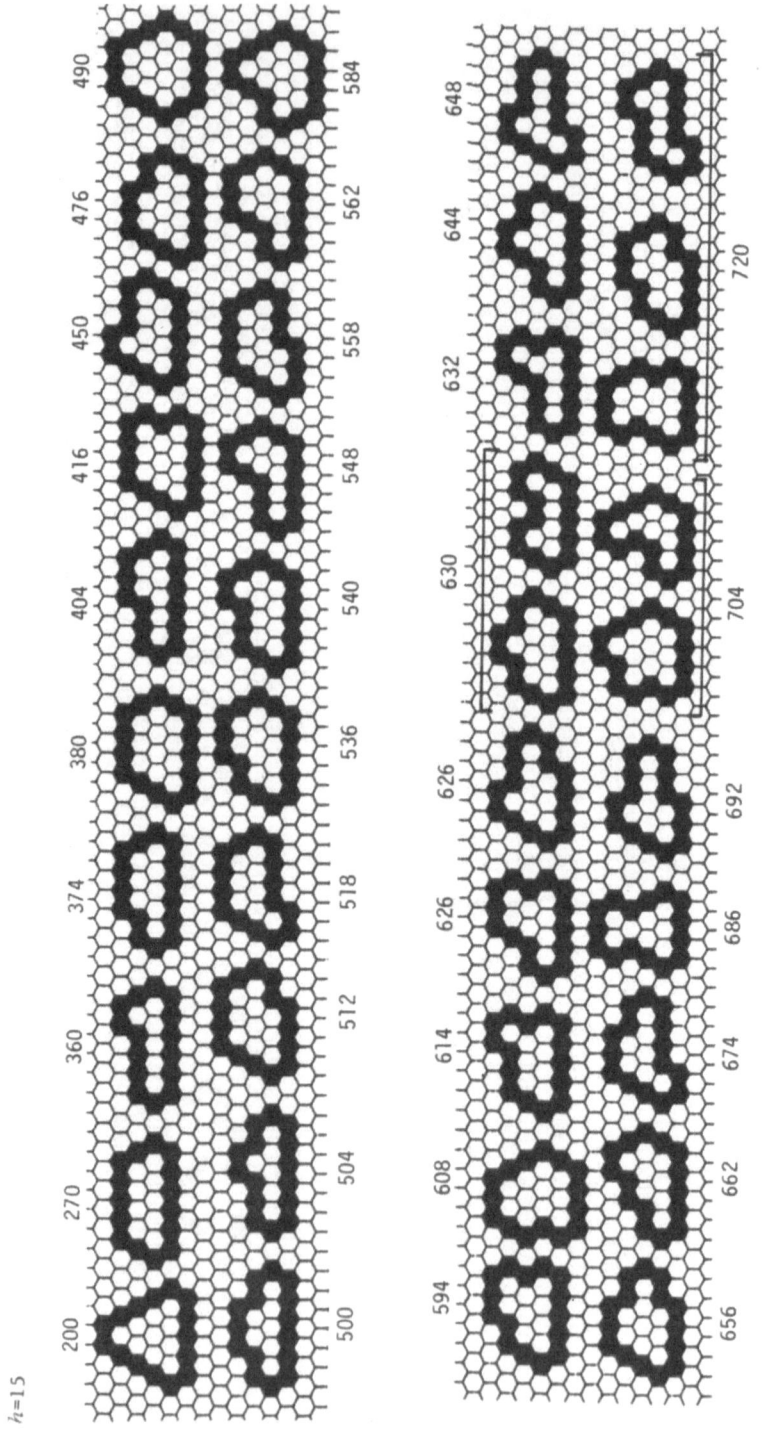

(cont.)

(Fig. 6.5 continued)

97

h=15 (continued)

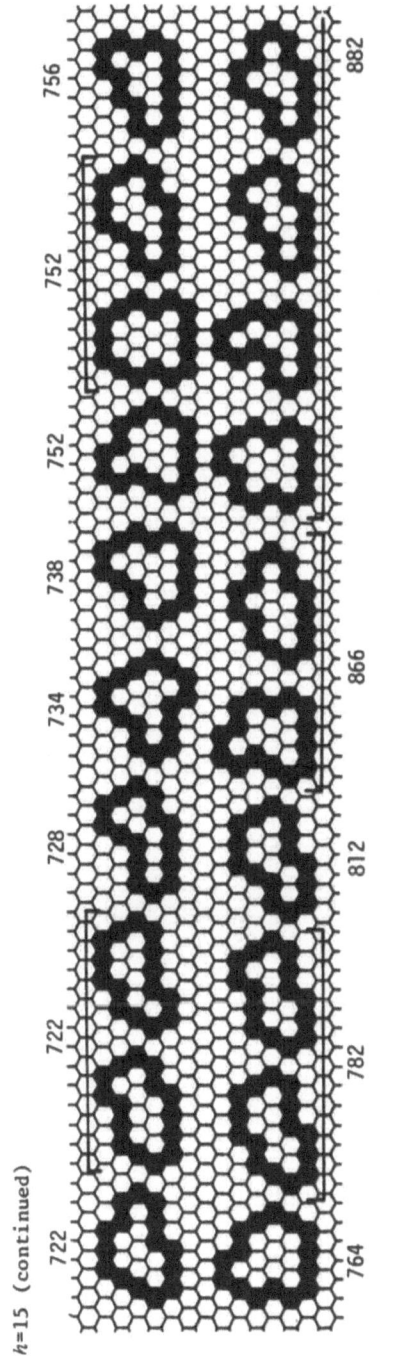

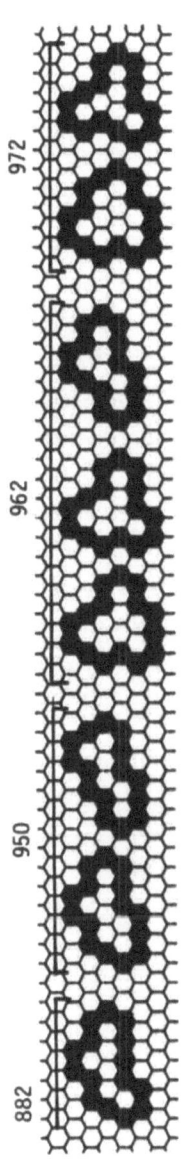

Fig. 6.5. All primitive coronoids for $8 < h \le 15$. K numbers are given. For the middle h=10 system Džonova-Jerman-Blažič and Trinajstić (1982) reported erroneously $K = 83$. For the first h=11 system Brunvoll, Cyvin BN and Cyvin (1987a) reported erroneously $K = 124$.

6.4 PRIMITIVE CORONOIDS WITH CONSTANT SEGMENT LENGTHS

6.4.1 *Introductory Remarks*

Consider a primitive coronoid $/l/^S$ in the notation of Section 4.3. It has S segments, each of the length l. The number of hexagons is

$$h = (l-1)S \tag{6.3}$$

Balaban et al. (1990) enumerated by computer aid some of the systems under consideration for $l=2$, i.e. for 2-segments. They also made some comments on the corresponding systems with $l > 2$ and especially $l=3$.

6.4.2 *Primitive Coronoids with 2-Segments Only*

The results of enumeration for the primitive coronoids $/2/^S$ for S up to 32 are shown in Table 6. The symmetry distribution is included.

The forms of the systems under consideration for $S = h \le 24$ are shown in Fig. 6. For each h value they are ordered according to the size of the corona hole. The systems with $h = 12$, 14 and 16 have $h^o = 4$, 6 and 8, respectively. For $h = 18$: $h^o = 7$ (1 system), $h^o = 10$ (3 systems), $h^o = 13$ (1 system). Further on, with numbers of systems indicated in parentheses: $h = 20$: $h^o = 9$ (1), $h^o = 12$ (3), $h^o = 15$ (1); $h = 22$: $h^o = 11$ (2), $h^o = 14$ (8), $h^o = 17$ (4), $h^o = 20$ (1); $h = 24$: $h^o = 10$ (1), $h^o = 13$ (4),

Table 6.6. Numbers of primitive coronoids with 2-segments only, classified according to symmetry.*

h	D_{6h}	D_{3h}	C_{3h}	D_{2h}	C_{2h}	C_{2v}	C_s	Total
12	1							1
14	0			1				1
16	0			0		1		1
18	1	1		1	1	1		5
20	0	0		0	0	3	2	5
22	0	0		4	2	4	5	15
24	0	2	1	0	0	6	20	29
26	0	0	0	5	9	19	48	81
28	0	0	0	0	0	22	159	181
30	2	4	3	9	29	48	422	517
32	0	0	0	0	0	59	1246	1305

* From: Balaban AT, Brunvoll J, Cyvin SJ (1990). Rev Roumaine Chim: in press

h^o = 16 (12), h^o = 19 (9), h^o = 22 (2), h^o = 25 (1).

In Table 6 there is no entry for the C_{6h} symmetry group. Such systems do exist, however, the smallest of them presumably having 42 hexagons. This system is depicted in Fig. 7.

It is an interesting fact that the corona hole of any primitive coronoid with 2-segments only is an all-benzenoid (Balaban et al. 1990).

6.4.3 *Primitive Coronoids with l-Segments Only, Especially for l > 2*

Primitive coronoids with l-segments only are possible for S = 6, 10, 12, 14, 16, ... when $l > 2$; for S = 12, 14, 16, ... when l=2.

It is not surprising that basically the same forms are encountered for primitive coronoids with constant segment lengths when l varies. The one-to-one correspondence is perfect for $l > 2$. Only for l=2 the numbers are reduced because of *coronene*, which is a quasi-coronoid, and the creation of helicenic systems, which also are quasi-coronoids. The top row of Fig. 8 shows $/3/^6$ and $/3/^{10}$. Their counterparts for l=2 are not coronoids. They are [6]*circulene* (*coronene* as a quasi-coronoid; cf. Fig. 1.4) and the helicenic system at the bottom of Fig. 2.6. For S = 12 there is a unique system of the considered type for every l. However, for S = 14 there are 3

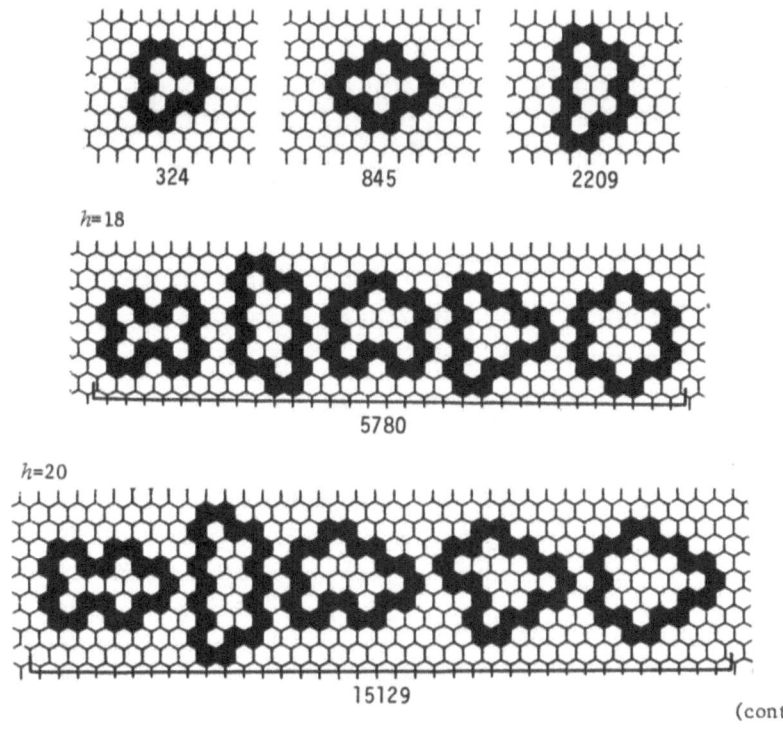

h=12 h=14 h=16

324   845   2209

h=18

5780

h=20

15129

(cont.)

(Fig. 6.6 continued)

(cont.)

(Fig. 6.6 continued)

h=24

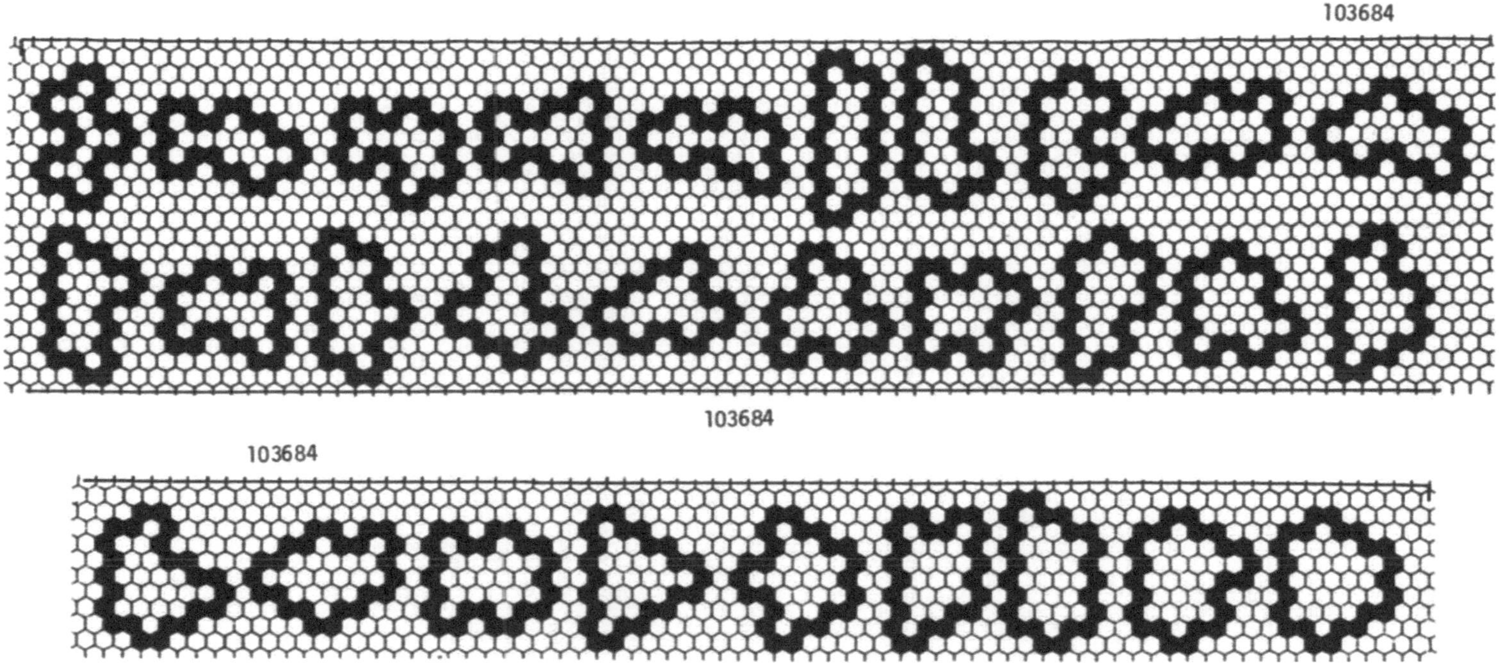

103684

103684

103684

Fig. 6.6. All primitive coronoids with 2-segments only for $12 \leq h \leq 24$, where h is an even number. K numbers (same value for all systems with the same h) are given.

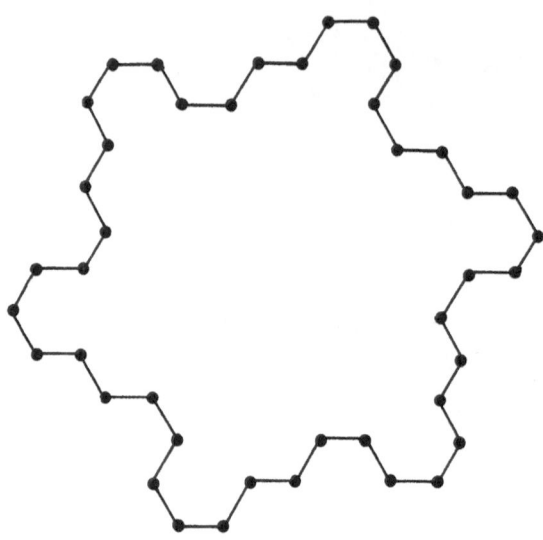

Fig. 6.7. The presumably smallest (h = 42) system with 2-segments only and C_{6h} symmetry.

systems for l > 2, two of them having helicenic counterparts when l=2; cf. Fig. 8. Further on ($S \geq 14$) there are constantly more $/l/^S$ systems for l > 2 than is the case for l=2.

Table 7 gives the numbers of primitive coronoids with constant segment lengths, $/l/^S$, when l > 2. The computations were carried out up to S = 26. Again here is no entry for the C_{6h} symmetry group. However, such forms exist, the presumably smallest one for a given l being the $/l/^{42}$ system which corresponds to Fig. 7.

Figure 9 shows the forms of primitive coronoids with 3-segments only up to S = 20 (h = 40). The systems with h = 12, 20 and 24 have h^o = 7, 15 and 25, respectively. For h = 28: h^o = 23 (2 systems), h^o = 35 (1 system); h = 32: h^o = 33 (1), h^o = 45 (1); h = 36: h^o = 31 (5), h^o = 43 (3), h^o = 55 (3), h^o = 67 (1); h = 40: h^o = 41 (6), h^o = 53 (4), h^o = 65 (3), h^o = 77 (1).

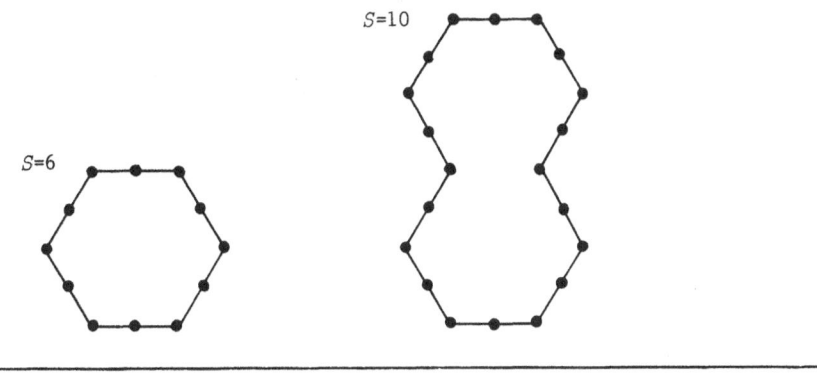

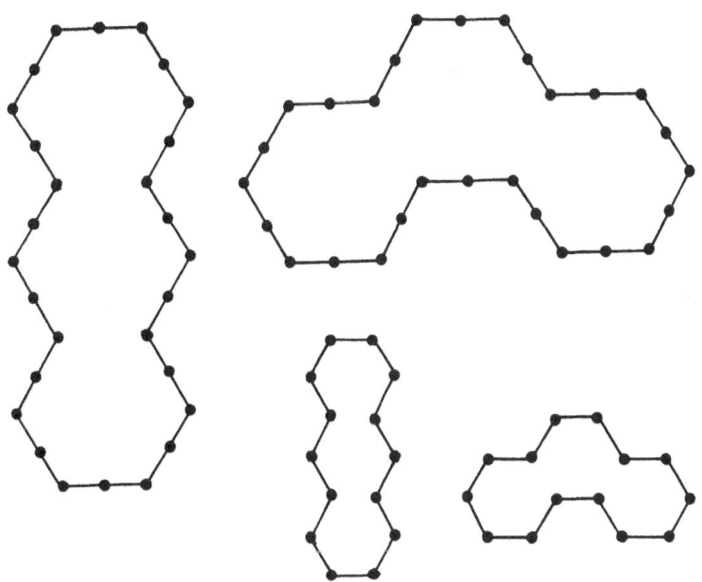

Fig. 6.8. The primitive coronoids $/3/^6$, $/3/^{10}$ and two of the $/3/^{14}$ forms as dualists. The helicenic quasi-coronoids with 2-segments corresponding to the two last forms are included.

Table 6.7. Numbers of primitive coronoids with l-segments only where $l > 2$, classified according to symmetry.

S	D_{6h}	D_{3h}	C_{3h}	D_{2h}	C_{2h}	C_{2v}	C_s	Total
6	1							1
10	0			1				1
12	0	1		0				1
14	0	0		2		1		3
16	0	0		0		1	1	2
18	1	2		2	2	3	2	12
20	0	0		0	0	5	9	14
22	0	0		6	5	14	25	50
24	0	2	2	0	0	9	84	97
26	0	0	0	9	20	52	231	312

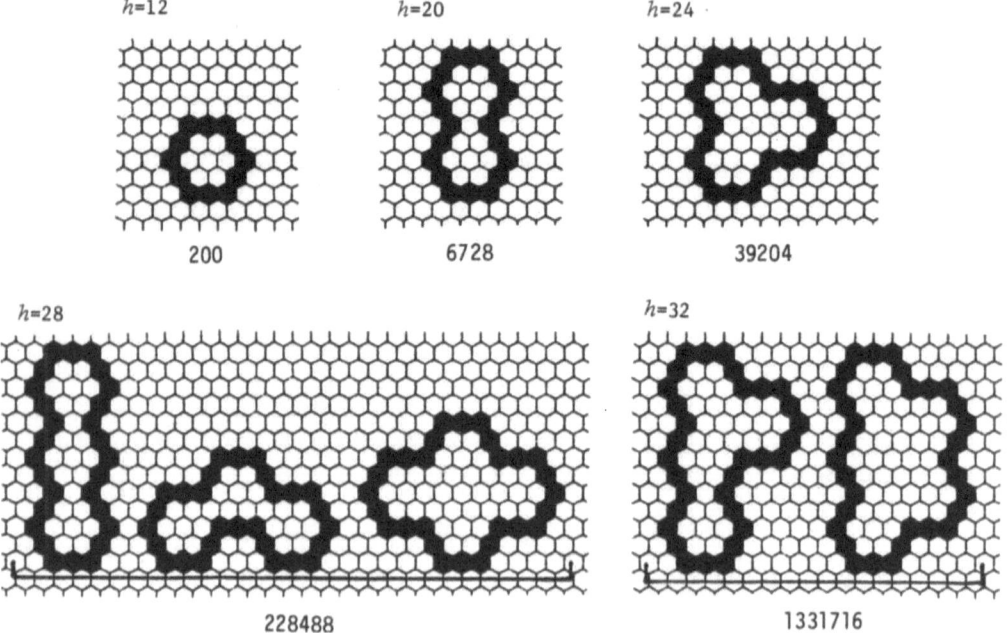

h=12 200

h=20 6728

h=24 39204

h=28 228488

h=32 1331716

(cont.)

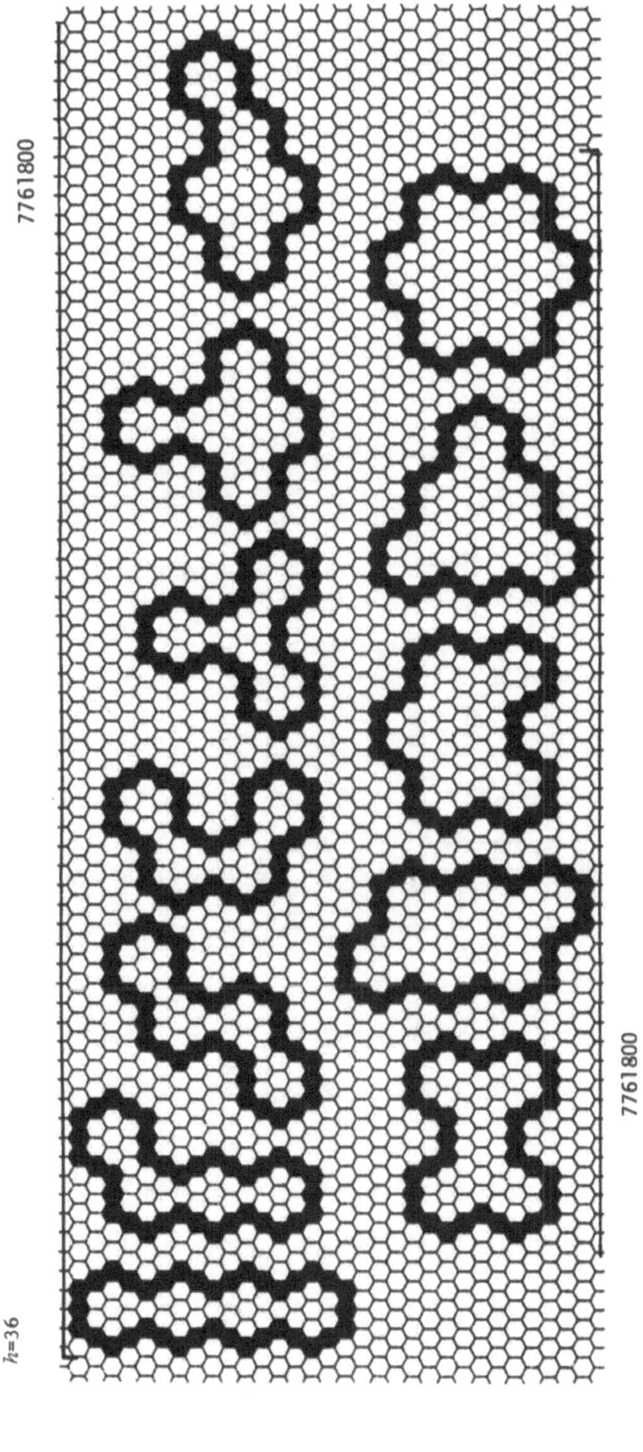

(Fig. 6.9 continued)

$h=36$

7761800

7761800

(cont.)

(Fig. 6.9 continued)

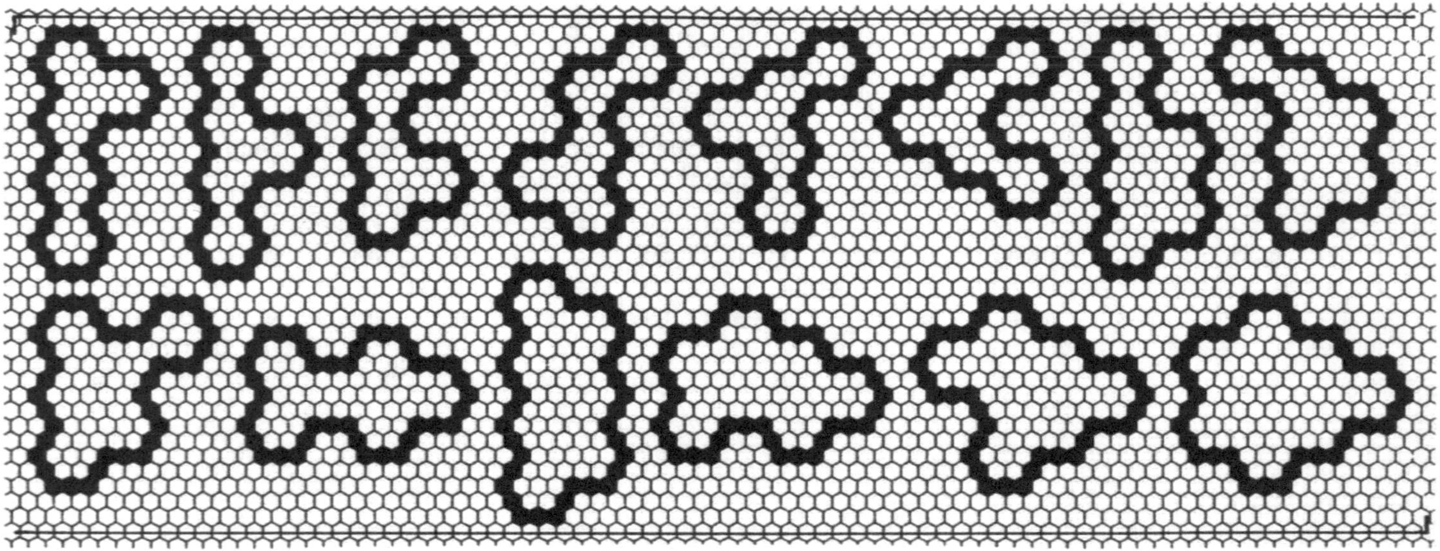

h=40

45239076

45239076

Fig. 6.9. All primitive coronoids with 3-segments only for $h = 12$ and $20 \leq h \leq 40$, where h is divisible by 4. K numbers (same value for all systems with the same h) are given.

6.5 ANNULENES

Here we report some results of enumeration for the [2*p*]*annulenes* according to their definition in Section 4.7. The topic has been treated elsewhere by Cyvin SJ, Brunvoll and Gutman (1990), who provided a substantial amount of supplements to the early work of Balaban (1971).

The [2*p*]*annulenes* of the considered type may be generated by a scheme very much similar to Fig. 4. The *p* values are correlated with the *h* values for the corresponding primitive coronoids according to eqn. (4.19). As a significant difference, the benzenoids associated with the *annulenes* may have coves and fjords in contrast to the corona holes of primitive coronoids. There are actually no restrictions for them. It was pointed out (Cyvin SJ, Brunvoll and Gutman 1990) that the enumeration of [2*p*]*annulenes* for a given *p* is equivalent to the enumeration of benzenoids with a given perimeter length. Therefore the extensive data of Stojmenović et al. (1986) may be transferred to the numbers of [2*p*]*annulenes* (see also Tošić et al. 1986). These numbers are collected in Table 8.

It is interesting to correlate the numbers of Tables 5 and 8. Each number for primitive coronoids with *h* hexagons (Table 5) should be compared with the number of [2*p*]*annulenes* (Table 8) with *p* = *h*-3; cf. eqn. (4.19). From the start we find the numbers to be identical. The first deviation occurs at *p*=9 (*h* = 12), where the *annulenes* count one more system. For higher *p* (and *h*) the difference is steadily increasing. The excess of *annulenes* is to be attributed exactly to the systems where the associated benzenoids have coves or fjords or otherwise are forbidden for the primitive coronoids.

Figure 10 shows some of the forms of [2*p*]*annulenes*.

Table 6.8. Numbers of [2p]annulenes.

p	$n = 2p$	Number
3	6	1^a
4	8	0^a
5	10	1^a
6	12	1^a
7	14	3^a
8	16	2^a
9	18	12^a
10	20	14^a
11	22	$50^{b,c}$
12	24	$97^{b,c}$
13	26	$312^{b,c}$
14	28	$744^{b,c}$
15	30	2291^c
16	32	6186^c
17	34	18714^c
18	36	53793^c
19	38	162565^c
20	40	482416^c
21	42	1467094^c
22	44	4436536^c
23	46	13594266^c

[a] Balaban AT (1971). Tetrahedron 27: 6115

[b] Tošić R, Doroslovački R, Gutman I (1986). Match 19: 219

[c] Stojmenović I, Tošić R, Doroslovački R (1986). Proceedings of the Sixth Yugoslav Seminar on Graph Theory, Dubrovnik 1985: 189

$2p = 6$ 10   12   14

  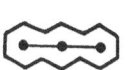 

$2p = 16$

$2p = 18$

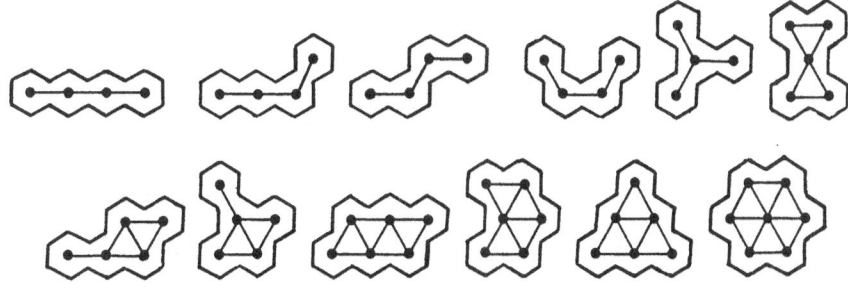

$2p = 20$

(cont.)

(Fig. 6.10 continued)

$2p = 22$

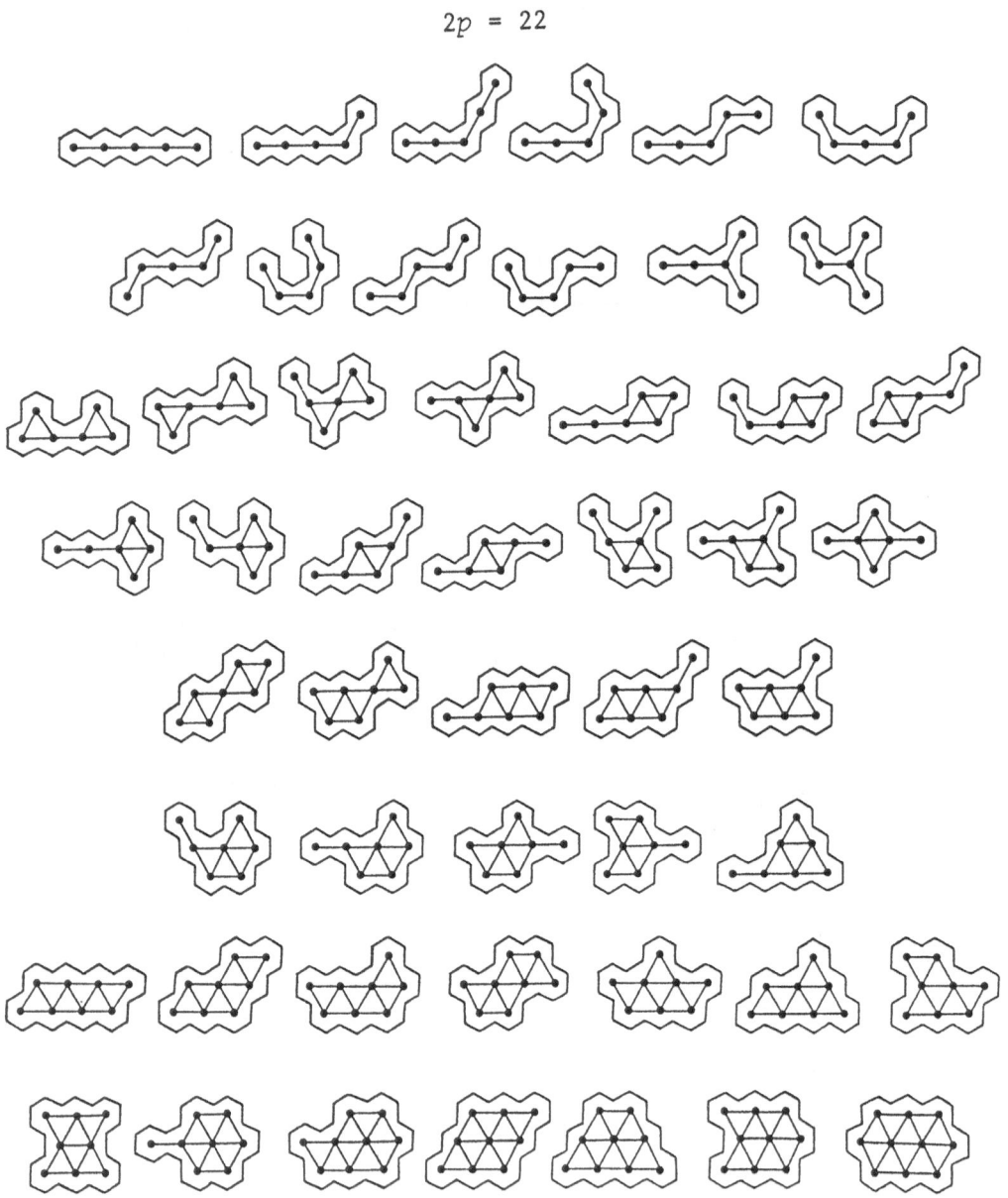

Fig. 6.10. The [2p]annulenes for $3 \leq p \leq 11$. The associated benzenoids are represented by dualists.

Chapter 7

ENUMERATION AND CLASSIFICATION OF NON-PRIMITIVE CORONOIDS

7.1 NON-PRIMITIVE KEKULÉAN CORONOIDS

Table 7.1. Numbers of non-primitive Kekuléan coronoids, classified according to symmetry.*

h	Type	C_{3h}	D_{2h}	C_{2h}	C_{2v}	C_s	Total
9	br	0	0	0	1	1	2
10	br	0	1	1	3	7	12
	rp	0	0	0	0	3	3
	he	0	0	2	2	2	6
11	br	0	0	0	11	49	60
	rp	0	0	0	2	26	28
	he	0	0	0	3	33	36
	e	0	0	0	0	2	2
12	br	1	4	7	25	264	301
	rp	0	0	3	9	202	214
	he	0	3	11	21	254	289
	e	0	1	2	1	35	39
13	br	0	0	0	58	1365	1423
	rp	0	0	0	28	1347	1375
	he	0	0	0	41	1780	1821
	e	0	0	0	6	413	419
14	br	0	9	43	140	6553	6745
	rp	0	2	22	90	8172	8286
	he	0	3	74	164	10947	11188
	e	0	5	24	21	3745	3795

*Abbreviations: br branched catacondensed; e essentially disconnected; he half essentially disconnected; rp regular pericondensed.

7.1.1 *Regular Systems*

The generation of regular coronoids is described in Paragraph 5.4.2. These systems are divided into classes according to the below scheme, which contains the presently applied abbreviations in parentheses.

$$
\text{regular } (r) \left\{
\begin{array}{l}
\text{unbranched catacondensed} \equiv \text{primitive } (prm) \\
\text{branched catacondensed } (br) \\
\text{regular pericondensed } (rp)
\end{array}
\right.
$$

The branched catacondensed systems are obtainable by specific generation: the primitive coronoids are subjected to one-contact additions. Consequently, when the numbers of primitive and total regular systems are known, the numbers of regular pericondensed coronoids are accessible through exclusion.

Table 1 (at the beginning of this chapter) includes the results of enumeration for non-primitive regular (br and rp) coronoid systems. The distribution into symmetry groups is given. The forms of all such systems ($br + rp$) for $9 \leq h \leq 11$ are displayed in Fig. 1.

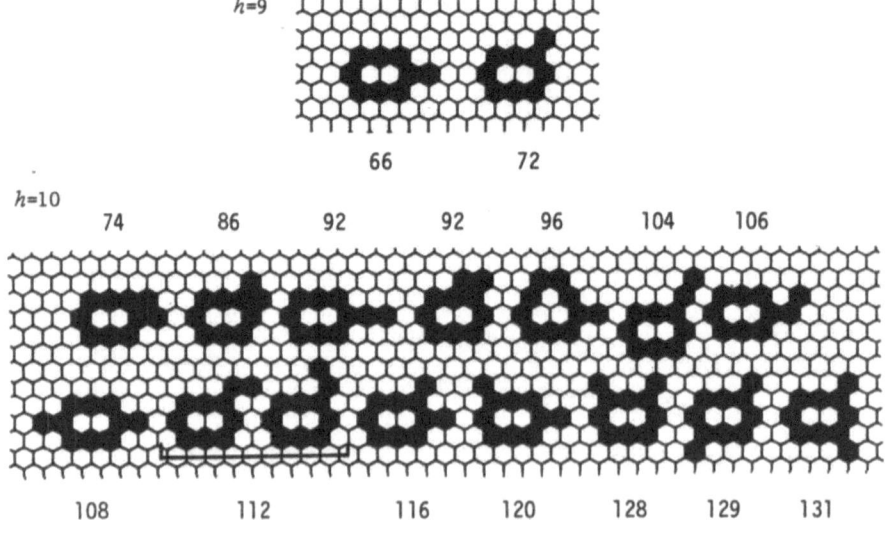

$h=9$

66   72

$h=10$

74   86   92   92   96   104   106

108   112   116   120   128   129   131

(cont.)

113

(Fig. 7.1 continued)

$h=11$

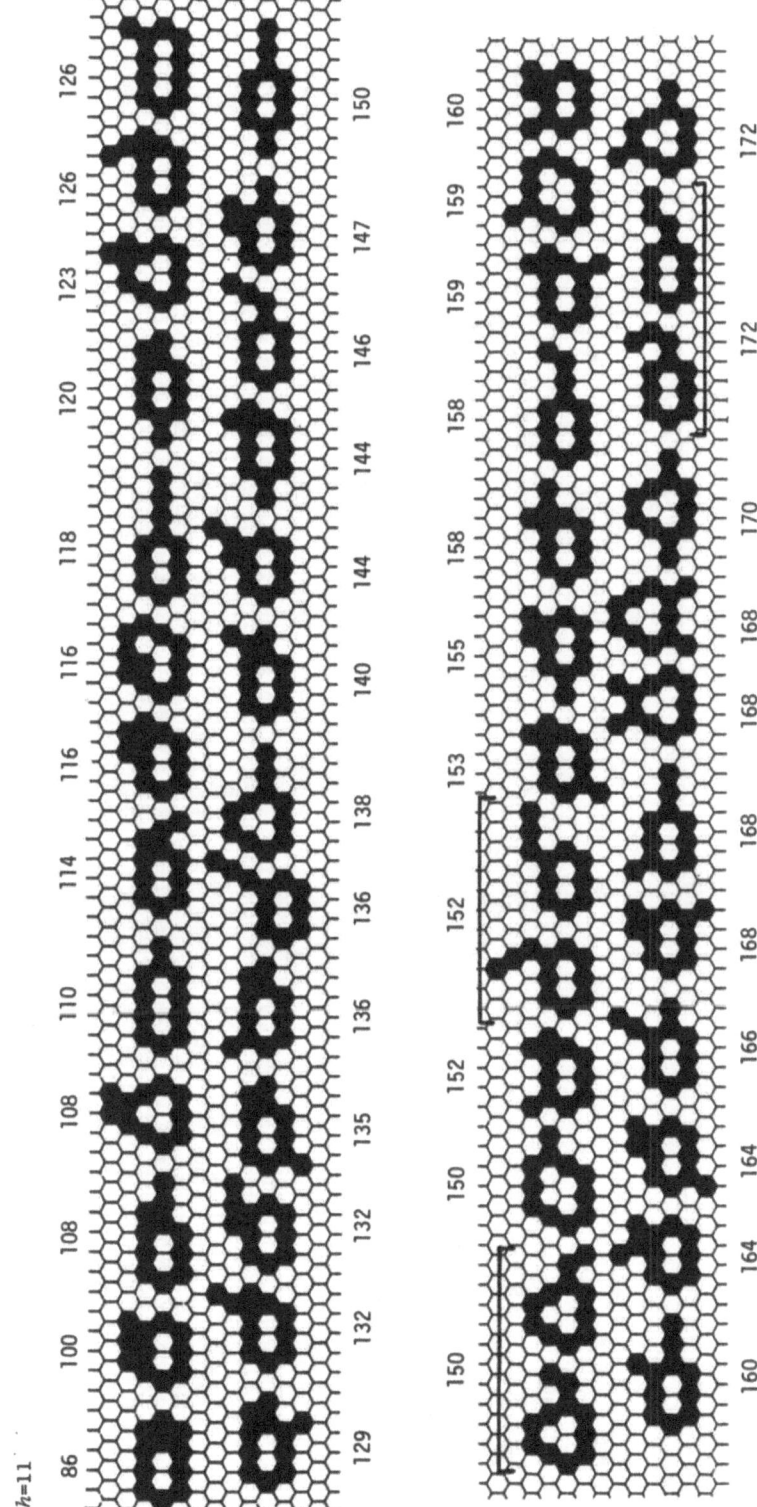

114

(Fig. 7.1 continued)

h=11 (continued)

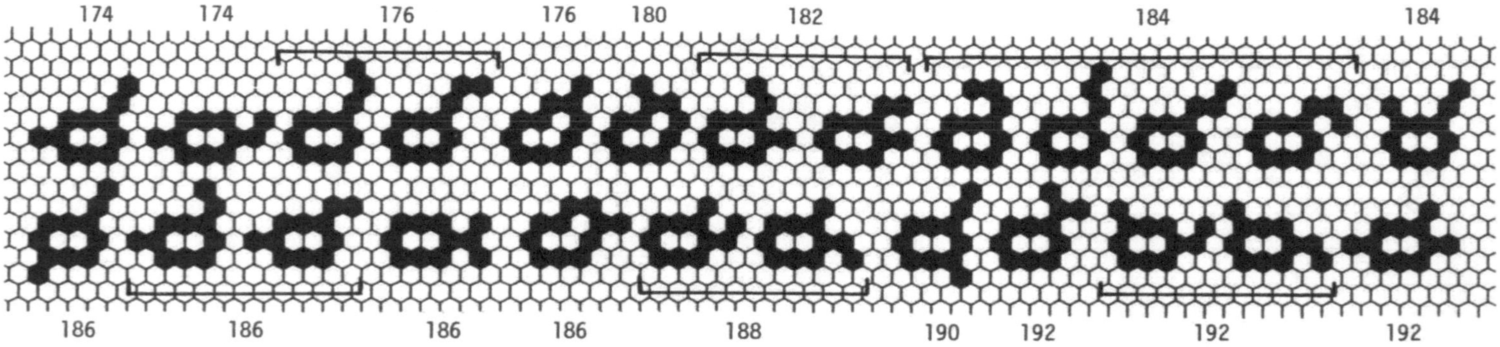

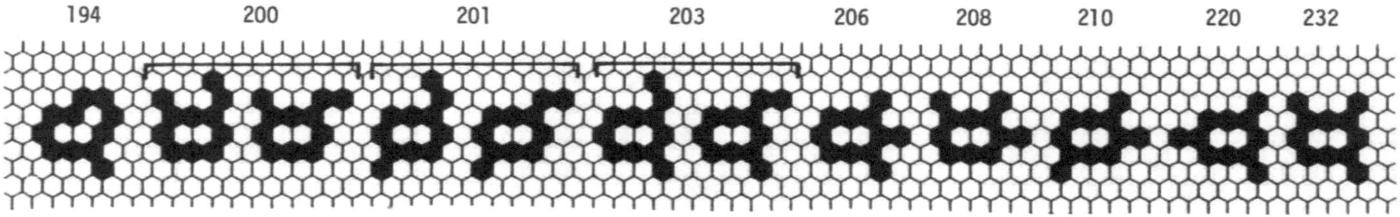

184   184   192

174 174 176   180 182   184   192 192

174   176   176   190

186   186   186   188   192

186

194   200   201   203   206   208   210   220   232

Fig. 7.1. All non-primitive regular coronoids for *h* ≤ 11. *K* numbers are given.

7.1.2 *Irregular Systems*

The irregular coronoid systems (see Paragraph 5.4.2) consist of the HED (half essentially disconnected, *he*) and essentially disconnected (*e*) coronoids.

Among the forms of relatively few members of a class of coronoids the essentially disconnected systems may often be recognized by sight. The method becomes impracticable if several hundred forms should be inspected for this purpose. For benzenoids, an authomatic method for recognizing essentially disconnected systems by computers has been described (Brunvoll, Cyvin SJ, Cyvin and Gutman 1989). It is based on the Pauling bond orders as obtained from the adjacency matrix. An adaptation to coronoids, where a skew-symmetric adjacency matrix is invoked (Cyvin SJ, Brunvoll and

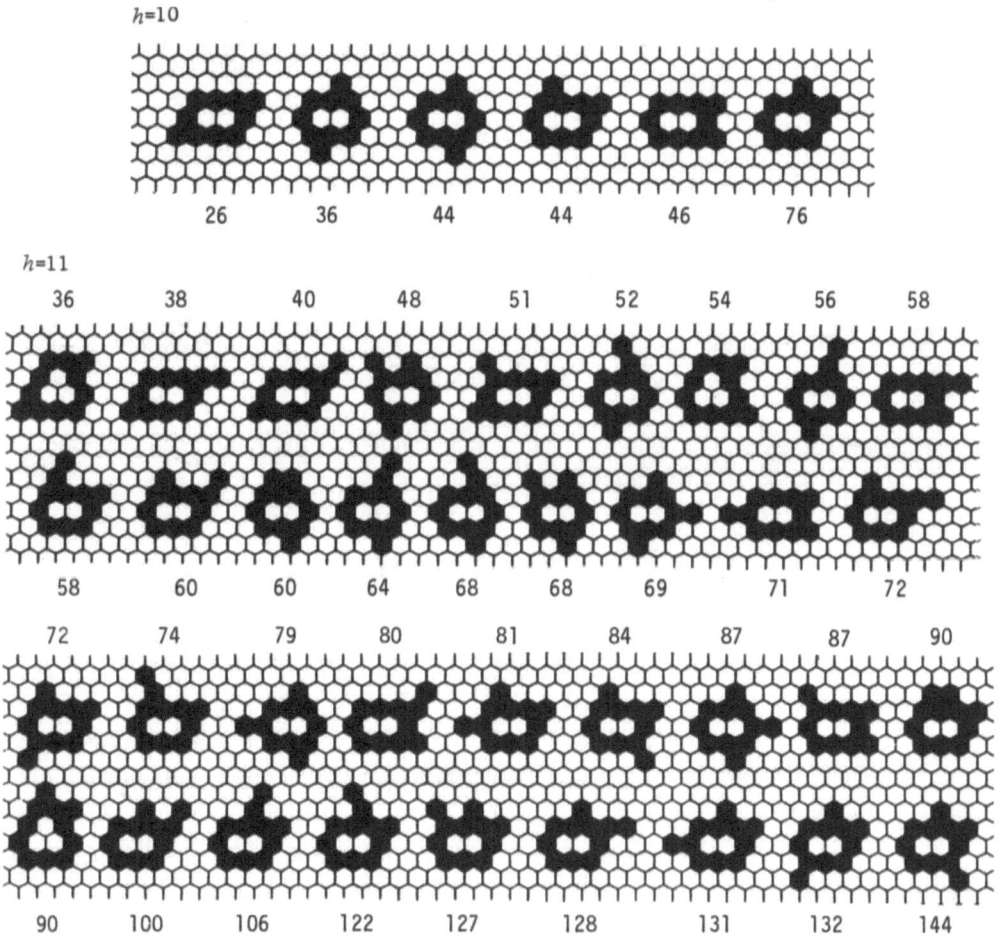

Fig. 7.2. All HED (half essentially disconnected) coronoids for $h \leq 11$. K numbers are given.

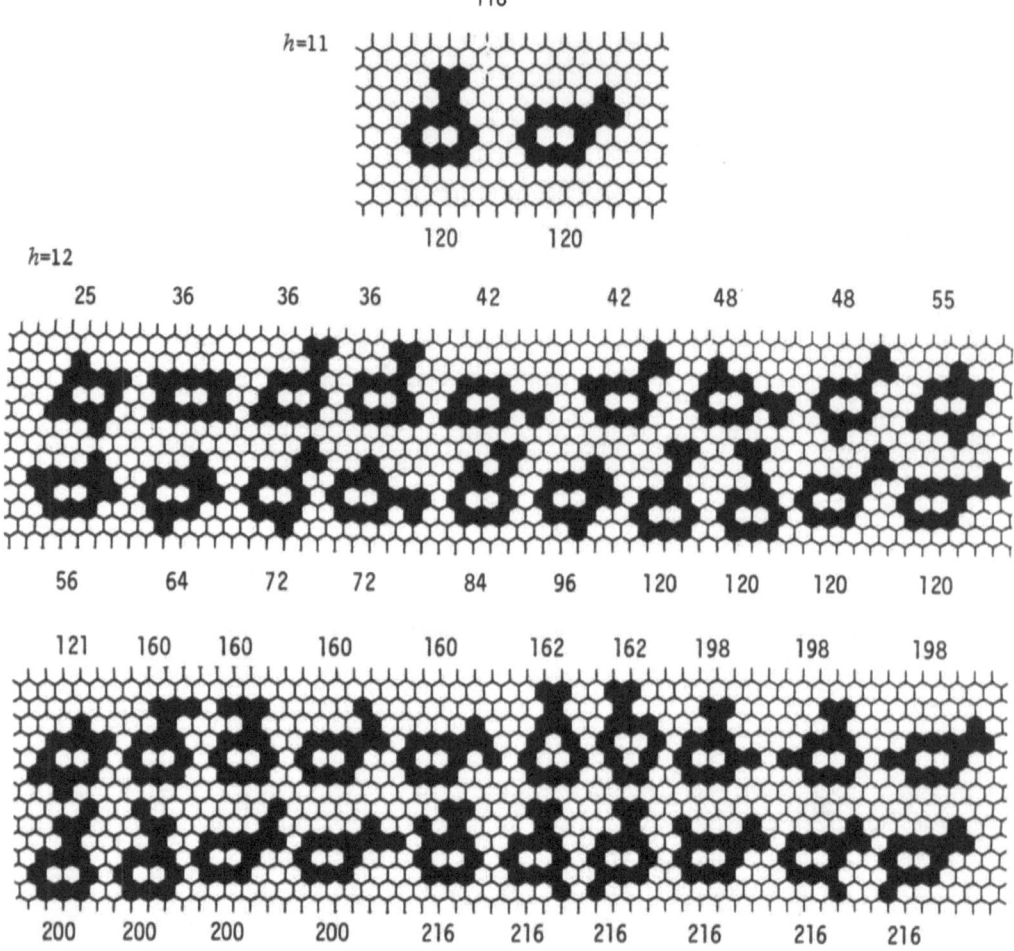

Fig. 7.3. All essentially disconnected coronoids for $h \leq 12$. K numbers are given.

Cyvin 1990c), has also been implemented.

In Balaban, Brunvoll et al. 1987 the reported number of essentially disconnected coronoids at $h = 12$ (tentatively determined by sight) is wrong; it should be 39;
cf. Table 1. The normal pericondensed (np) systems are found on summation ($rp + he$).
Also this number (np) for $h = 12$ was given erroneously in Balaban, Brunvoll et al.
1987; it should be 503. However, the sum ($np + e$), viz. the number of pericondensed
Kekuléan coronoids was given correctly as 542. On the other hand, the number of pericondensed Kekuléan coronoids with $h = 14$ is reported with one unit off in He WJ et
al. 1988; the correct number should be 23269.

The results of enumeration of the irregular coronoid systems, classified
according to symmetry, are included in Table 1. Figure 2 shows the forms of all HED
coronoid systems with $h = 10$ and $h = 11$, while the forms of all essentially disconnected coronoids with $h = 11$ and $h = 12$ are displayed in Fig. 3.

7.2 CLASSIFICATION ACCORDING TO THE COLOR EXCESS

The enumeration of coronoids with different Δ values (color excess) was treated in some detail in Paragraph 5.4.3. Table 2 shows the numbers of coronoids with
the different Δ values within each symmetry group. The sums of the numbers added
horizontally agree with those of Table 5.4. The sums added vertically agree with
the numbers of Table 5.5. Table 2 is complete up to $h = 14$. Incomplete results for
higher h values and relatively large Δ, consistent with entries in Table 5.4, are
collected in Table 3.

Table 7.2. Numbers of coronoids with different color excess (Δ), classified according to symmetry: complete list for $h \leq 14$.*

h	Δ	D_{6h}	D_{3h}	C_{3h}	D_{2h}	C_{2h}	C_{2v}	C_s
8	0	0	0	0	1	0	0	0
9	0	0	1	0	0	0	1	1
	1	0	0	0	0	0	0	2
10	0	0	0	0	3	3	6	12
	1	0	0	0	0	0	2	14
	2	0	0	0	0	0	1	2
11	0	0	0	0	0	0	17	111
	1	0	0	0	0	0	3	128
	2	0	0	0	0	0	4	19
	3	0	0	0	0	0	0	1
12	0	1	2	1	10	25	58	757
	1	0	0	0	0	0	12	894
	2	0	0	0	0	0	8	174
	3	0	1	1	0	0	1	9
13	0	0	0	0	0	0	137	4913
	1	0	0	0	0	0	24	5889
	2	0	0	0	0	0	18	1283
	3	0	0	0	0	0	5	93
	4	0	0	0	0	0	1	0
14	0	0	0	0	25	168	427	29434
	1	0	0	0	0	0	77	36478
	2	0	0	0	0	0	63	8812
	3	0	0	0	0	0	9	776
	4	0	0	0	0	0	3	11

*Systems of C_{6h} symmetry occur for $h \geq 18$.

Table 7.3. Numbers of coronoids with different color excess (Δ), classified according to symmetry: incomplete list for $h > 14$.

h	Δ	D_{3h}	C_{3h}	C_{2v}	C_s
15	2	0	0	121	57384
	3	3	7	33	5952
	4	0	0	5	158
16	3	0	0	73	42915
	4	0	0	25	1586
	5	0	0	0	4
17	4	0	0	49	14267
	5	0	0	5	110
18	5	0	0	9	1602
	6	2	0	0	0
19	6	0	0	3	29

The coronoid systems with $\Delta=0$ are all Kekuléan for $h \leq 14$; this is as far as
Table 2 goes. Such systems are treated above (Chapter 6 and Section 7.1). All systems with $\Delta > 0$ are by definition obvious non-Kekuléan (cf. Paragraph 3.3.3). Figure
4 shows the forms of all obvious non-Kekuléan coronoids with $h = 9$ and $h = 10$. The
systems with $\Delta=2$ and $\Delta=3$ for $h = 11$ are depicted in Fig. 5.

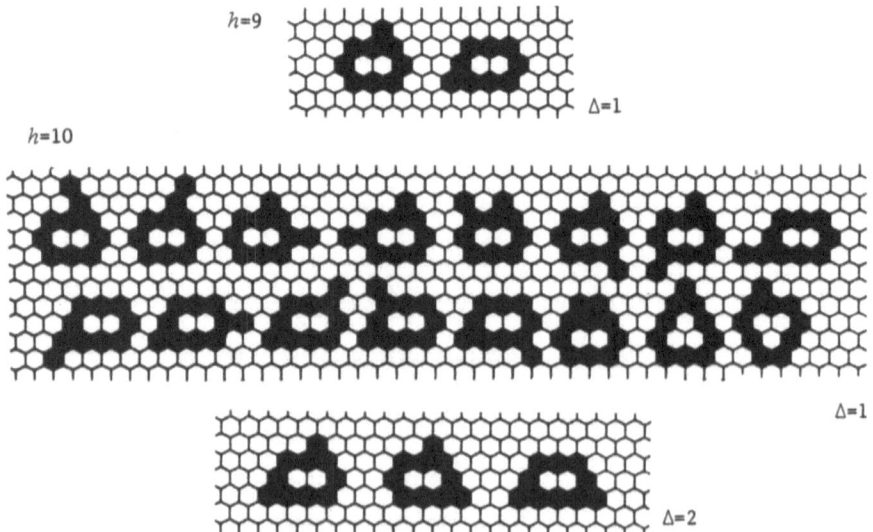

Fig. 7.4. All non-Kekuléan coronoids for $h \leq 10$; they are obvious non-Kekuléan.

h=11

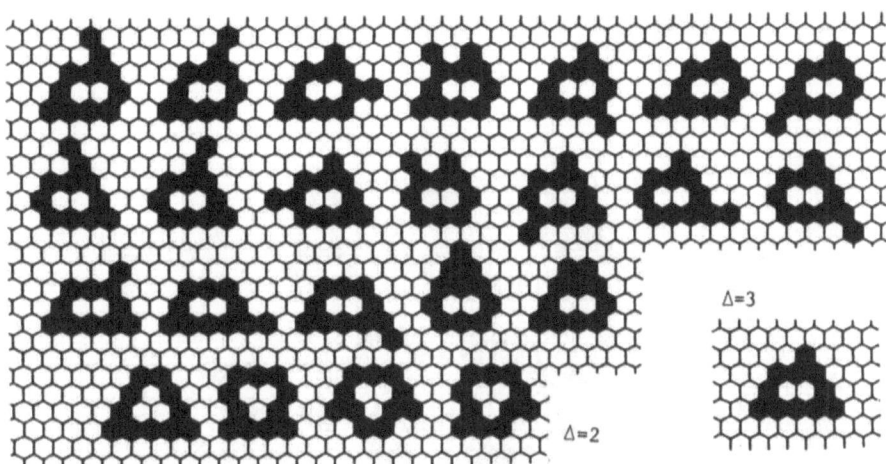

Fig. 7.5. All (obvious non-Kekuléan) coronoids with Δ=2 and Δ=3 for h = 11.

Especially for the extremal obvious non-Kekuléan coronoids (with $\Delta = \Delta_{max}$),
see Paragraph 5.4.3.

7.3 CONCEALED NON-KEKULÉAN CORONOIDS

The very plausible assumption was made that the smallest concealed non-Kekuléan coronoid system (K=0, Δ=0; cf. Paragraph 3.3.3) has a *naphthalene* (h° = 2)
corona hole. It was demonstrated that the smallest system of this category has h =
15, provided that the above assumption is right. The procedure is described below.

A complete generation and classification of the h = 14 coronoids with *naphtha-*
lene hole was performed (see the next chapter) as outlined in the preceding sections. It was ascertained that no concealed non-Kekuléan coronoid exists among these
56003 systems. By virtue of the selection rules for Δ (Paragraph 5.4.3) all the (unknown number of) Δ=0 systems with h = 15 may be generated with certainty by additions of hexagons to the 22154 systems of h = 14, Δ=0 and the 26919 systems of h =
14, Δ=1. The principle of sifting (Section 5.2) was applied in such a way that only
the systems with K=0 (within the class of those with h = 15, Δ=0) were retained,
while all the others were discarded. The result of this analysis are the 23 systems
depicted in Fig. 6. They all belong to the C_{s} symmetry.

The assumption about *naphthalene* hole was proved by further computer-enumera-
tions. The coronoids with *phenalene* hole were enumerated (see the next chapter) with
a detailed classification for all such systems with $h \leq$ 15. The analysis confirmed

that there are no concealed non-Kekuléans among these coronoids. Likewise was the non-existence of concealed non-Kekuléans with h = 15 proved for all the other thinkable holes, viz. those which at all are compatible with $h \leq 15$ coronoids.

h=15

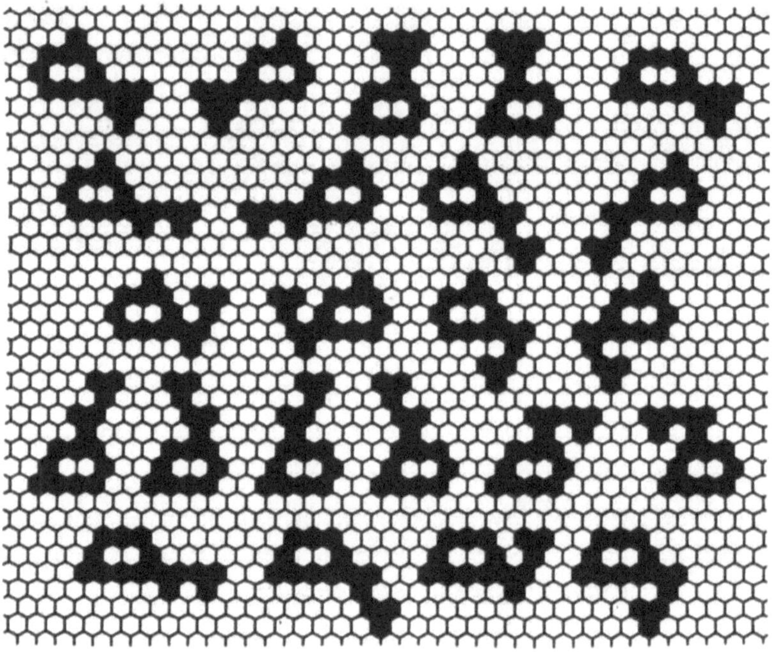

Fig. 7.6. The smallest concealed non-Kekuléan coronoids (with a hole of h^o = 2); they have h = 15.

Chapter 8

ENUMERATION AND CLASSIFICATION OF CORONOIDS WITH DEFINITE HOLES

8.1 INTRODUCTION

A classification of coronoids according to the corona hole is especially in-
structive. Firstly, we shall be concerned about the enumeration of all coronoids
with a given hole and different (increasing) numbers of hexagons (h). Secondly, a
detailed subdivision into classes within the whole set shall be reported for the
different $\overset{\prime}{h}$ values.

In preparation of these enumerations and classifications some supplementary
studies of topological properties of coronoids are warranted, namely those of particular relevance to the shapes of the corona holes. The preceding chapters on ANATOMY
are of interest in this context: Chapter 3, and even more Chapter 4.

8.2 SUPPLEMENTARY TOPOLOGICAL PROPERTIES

8.2.1 *Outside and Inside Features*

In any coronoid system a smallest primitive coronoid which is part of the system, can uniquely be distinguished. If the system is not a primitive coronoid itself,
it must have either an *outside feature* or an *inside feature* (or both). The outside
features may be generated by adding hexagons to the outer perimeter of the primitive
coronoid. Allowance is made for repeated additions so that the hexagons of the outside features are (directly or indirectly) connected with the outer perimeter. Cor-
respondingly the inside features (Brunvoll, Cyvin BN and Cyvin 1987a) may be generated by adding hexagons to the inner perimeter.

In Fig. 3.4 the coronoid has clearly one outside and one inside feature, consisting of one hexagon each. Below we show a more advanced example.

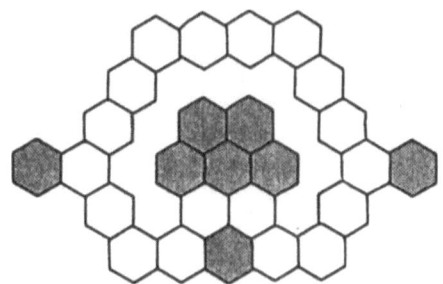

Here the three grey isolated hexagons represent the outside features. The inside feature consists of the five grey connected hexagons. Many other examples are found through Chapter 3 and elsewhere in this book.

8.2.2 *Basic Coronoid*

Every corona hole (defined by a benzenoid) determines uniquely a *basic coronoid* (system). That is the smallest coronoid with this hole.

All primitive coronoids are basic coronoids. But, as was mentioned in Paragraph 4.2.2, there exist benzenoids which can not serve as corona holes for primitive coronoids. The corresponding basic coronoids have inside features, but never an outside feature.

Figure 1 shows four basic coronoids which are not primitive coronoids; we shall refer to such systems as *non-primitive basic coronoids*. The same corona holes are found in Fig. 4.1. These four systems re-appear among the systematic listing of non-primitive basic coronoids (see below).

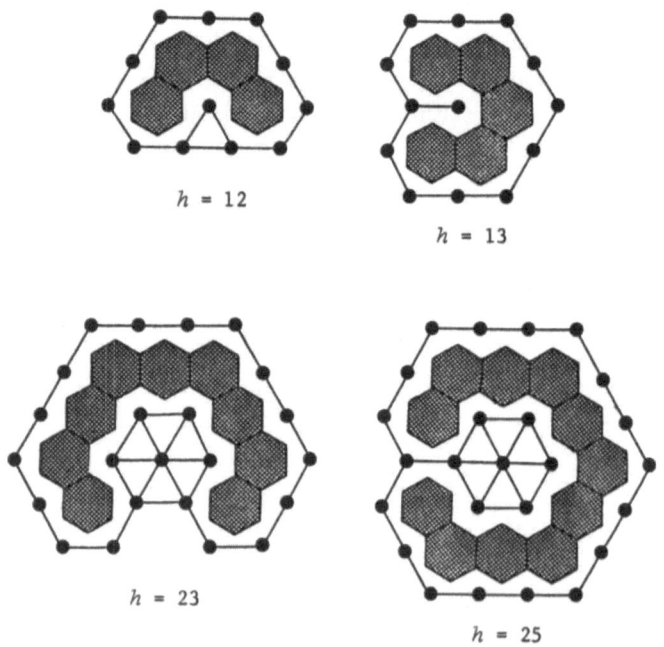

$h = 12$

$h = 13$

$h = 23$

$h = 25$

Fig. 8.1. Four non-primitive basic coronoid systems as dualists. Dark hexagons show the corona holes. See also Fig. 4.1.

8.3 BASIC CORONOIDS

8.3.1 *Generation and Enumeration*

The listing of basic coronoid systems is the same as a listing of the corona holes. For the generation of the corona holes a method similar to the one of Paragraph 6.3.2 for primitive coronoids was employed. In the present case there are no restrictions (apart from $h^o > 1$) on the benzenoids which represent the corona holes.

Another principle for generating basic coronoid systems is based on the following property. All non-primitive basic coronoids with $h+1$ hexagons are obtained by adding (when possible) one hexagon every time to the inner perimeters in selected positions of all basic (primitive and non-primitive) coronoids with h hexagons.

Table 1 gives the result of an enumeration of basic coronoid systems without further classification. It is noted that h designates the number of hexagons of the basic coronoid, not the corona hole.

The stated rule about generation of basic coronoids ($h \rightarrow h+1$) makes it clear that the maximum value of Δ, viz. Δ_{max}, can not increase by more than one unit when h increases by one; cf. the selection rules for Δ in Paragraph 5.4.3. Table 1 includes the Δ_{max} values for $h \leq 18$.

Table 8.1. Numbers of basic coronoid systems and their maximum Δ value.

h	Number	Δ_{max}
8	1	0
9	1	0
10	3	0
11	2	0
12	12	1
13	15	1
14	51	1
15	109	1
16	343	2
17	903	2
18	2821	3

Table 8.2. Numbers of non-primitive basic coronoid systems, classified according to Δ and symmetry.*

h	Δ	D_{3h}	C_{3h}	D_{2h}	C_{2h}	C_{2v}	C_s	Total
12	1	0	0	0	0	1	0	1
13	0	0	0	0	0	1	0	1
	1	0	0	0	0	1	1	2
14	0	0	0	0	0	0	2	2
	1	0	0	0	0	1	8	9
15	0	0	0	0	0	4	9	13
	1	0	0	0	0	3	25	28
16	0	0	0	2	2	4	39	47
	1	0	0	0	0	6	95	101
	2	0	0	0	0	1	2	3
17	0	0	0	0	0	16	155	171
	1	0	0	0	0	8	315	323
	2	0	0	0	0	4	10	14
18	0	0	0	1	18	32	569	620
	1	0	0	0	0	11	1059	1070
	2	0	0	0	0	6	62	68
	3	1	1	0	0	0	0	2

*Also hexagonal symmetries occur for higher h values.

The numbers of Table 1 incorporate the primitive coronoids, which are enumerated specifically in Table 6.5. The non-primitive basic coronoid systems are simply obtained by exclusion. It is found that the smallest non-primitive basic coronoid is a unique system for $h = 12$. The results up to $h = 18$ are collected in Table 2, where the systems under consideration are classified according to their color excess (Δ) and the symmetry groups.

Up to $h = 16$ there are more primitive coronoids than the non-primitive basic coronoids for a given h, but the relative difference decreases with increasing h. For $h \geq 17$ the magnitudes have switched, and the non-primitive basic coronoids dominate more and more with increasing h.

8.3.2 Forms

Figure 2 displays the non-primitive basic coronoid systems for $h = 12, 13, 14$ and 15. In all but one case each system has an inside feature consisting of one hexagon only. Hence a system with h hexagons is obtained by adding one hexagon into the corona hole of a primitive coronoid with $h-1$ hexagons. The exception is the first system with $h = 15$, $\Delta=1$. Here the inside feature consists of two hexagons;

the basic coronoid is obtained by adding these two hexagons to an $h = 13$ primitive coronoid. The obvious non-Kekuléan systems ($\Delta=1$) in Fig. 2 are listed in the same order as the respective primitive coronoids in Fig. 6.5. The Kekuléan systems ($\Delta=0$), on the other hand, are ordered according to increasing K numbers.

For the sake of further illustration we give the forms of some of the smallest extremal basic coronoid systems. Figure 3 shows the systems with $h = 16$ and $h = 17$ possessing $\Delta = \Delta_{max} = 2$. The smallest systems with $\Delta=3$ ($h = 18$ and $h = 19$) are displayed in Fig. 4.

It was stated in Paragraph 4.2.2 that the corona hole of a primitive coronoid can not have a cove or fjord. All the corona holes of the non-primitive basic coronoids in Fig. 2 possess either a cove or a fjord. Similarly the corona holes of the systems in Fig. 3 have two coves each and those in Fig. 4 three coves each. But it was also stated (Paragraph 4.2.2) that benzenoids without any cove or fjord exist, which nevertheless can not be used as corona holes for primitive coronoids. The above material is not large enough to provide an example. The smallest basic coro-

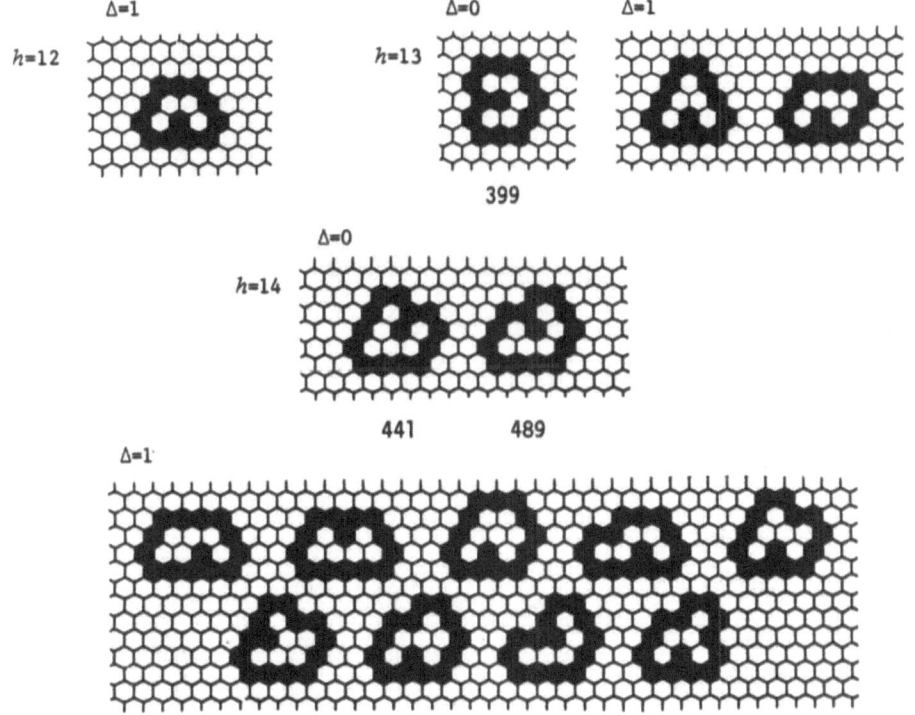

(cont.)

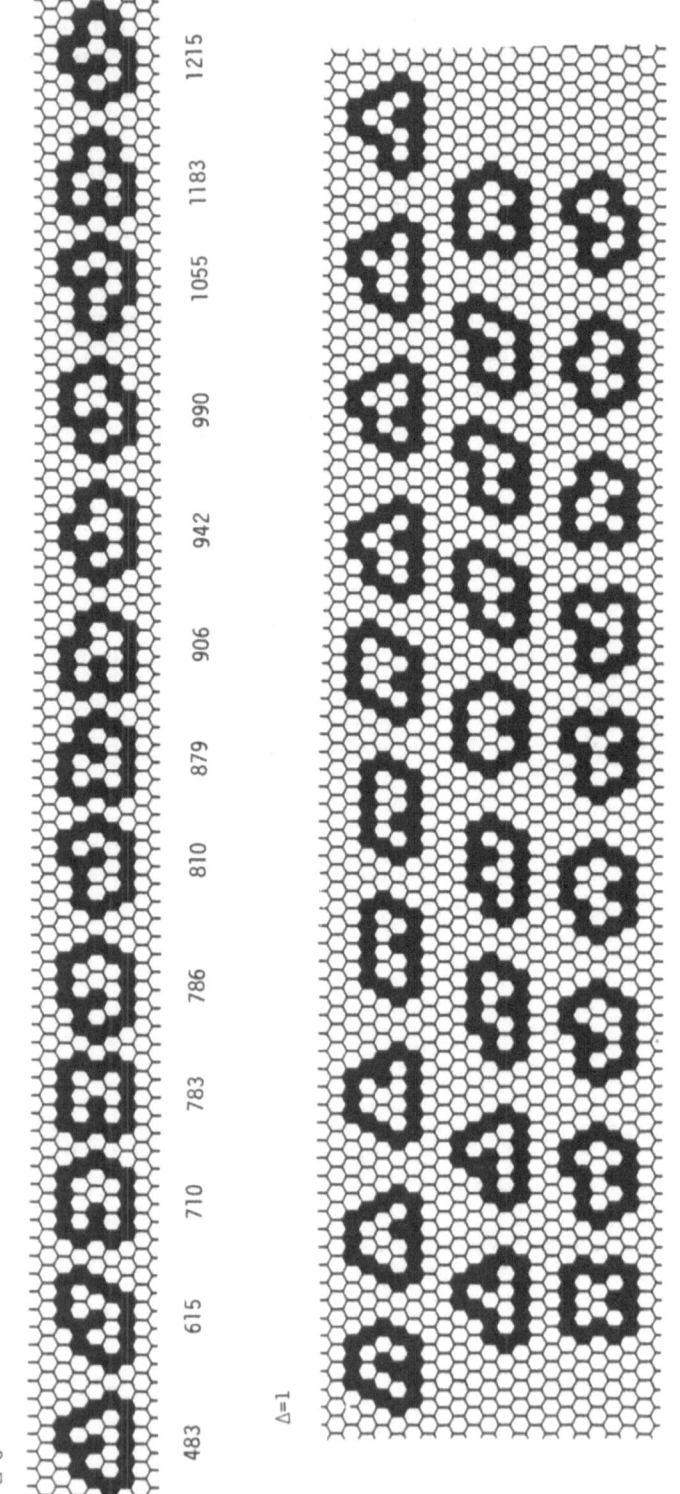

(Fig. 8.2 continued)

$h=15$

$\Delta=0$

483 615 710 783 786 810 879 906 942 990 1055 1183 1215

$\Delta=1$

Fig. 8.2. All non-primitive basic coronoid systems for $h \leq 15$. K numbers are given for the Kekuléan coronoids.

h=16

Δ=2

h=17

Δ=2

Fig. 8.3. The smallest basic coronoid systems with Δ=2.

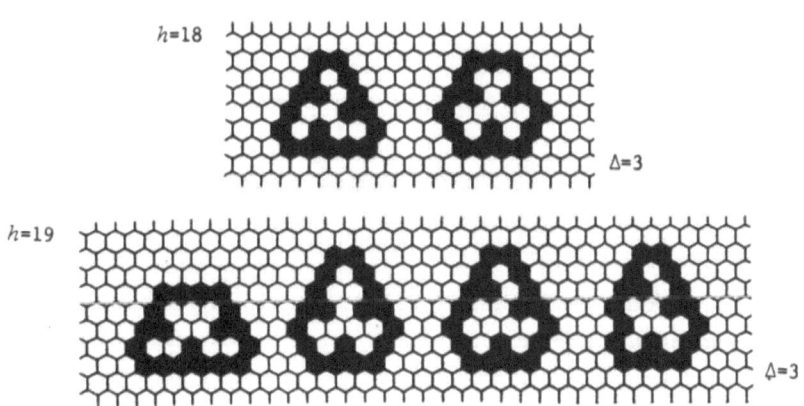

h=18

Δ=3

h=19

Δ=3

Fig. 8.4. The smallest basic coronoid systems with Δ=3.

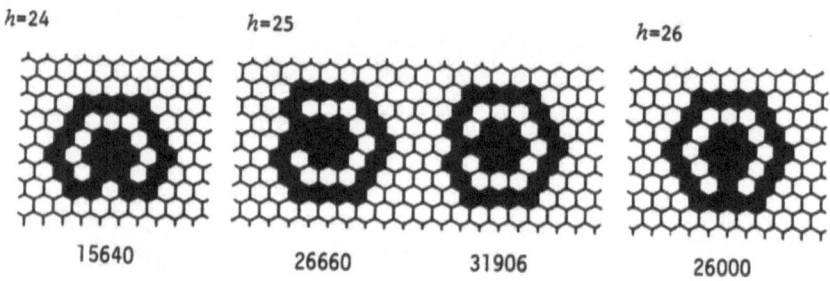

Fig. 8.5. The presumably smallest basic coronoid systems where the corona holes have no cove and no fjord. In all these cases, Δ=0. The K numbers are given.

noid with such a hole, which was found by trial and error, is a unique system with h = 23; see Figs. 1 and 5. These figures also show a system of this category for h = 25. Figure 5 includes one additional, less symmetrical system with the same number of hexagons, and finally one system with h = 26.

8.4 DEFINITE CORONA HOLES

8.4.1 *First Survey*

Table 3 shows the corona holes up to h=12, where h is the number of hexagons of the corresponding basic coronoids. Each of these basic coronoids was used as the starting point for specific generation of all the coronoids with increasing h values and possessing the definite hole. The results for $h_{basic} \leq 12$ are entered in Table 3. Here h_{basic} designates the number of hexagons of the basic coronoid in question.

8.4.2 *Second Survey*

The same computations (above paragraph) were pursued to h_{basic} = 16. Table 4 gives the results from this more extensive material, but presented in a more compressed form. The data from all holes with the same h_{basic} value have been added together.

In Tables 3 and 4 the figure for h = 15 systems with *naphthalene* hole was determined by exclusion with the aid of the grand total (Table 5.1). Therefore the full details on classification for this class are unknown.

Table 8.3. Numbers of coronoids with different corona holes (first survey).

CORONA HOLE

h									
8	1[a]								
9	4[a]	1[a]							
10	37[a]	3[a]	1[a]	1[a]	1[a]				
11	235[a]	27[a]	5[a]	5[a]	9[a]	1[a]	1[a]		
12	1554	186	46	46	84	9	17	1	1
13	9450	1245	323	323	636	92	176	9	5
14	56003	7803	2247	2247	4421	731	1449	92	56
15	321409	47116	14396	14396	28479	5279	10458	731	423

CORONA HOLE

h										
12	1	1	1	1	1	1	1	1	1	1
13	4	9	18	18	2	10	9	5	10	4
14	37	102	195	195	22	103	102	56	103	38
15	289	834	1665	1665	146	843	831	421	842	289

[a]Brunvoll J, Cyvin BN, Cyvin SJ (1987). J Chem Inf Comput Sci 27: 14

Table 8.4. Numbers of coronoids with different corona holes (second survey).

h	h value of the basic coronoid (h_{basic})							
	8	9	10	11	12	13	14	15
8	1^a							
9	4^a	1^a						
10	37^a	3^a	3^a					
11	235^a	27^a	19^a	2^a				
12	1554	186	176	26	12			
13	9450	1245	1282	268	103	15		
14	56003	7803	8915	2180	1101	230	51	
15	321409	47116	57271	15737	8979	2568	757	109

[a] Brunvoll J, Cyvin BN, Cyvin SJ (1987). J Chem Inf Comput Sci 27: 14

Table 8.5. Enumeration and detailed classification of coronoids with the *naphthalene* hole.*

CORONA HOLE

h	Type	Δ	D_{2h}	C_{2h}	C_{2v}	C_s	Total
8	*prm*	0	1	0	0	0	1
9	*br*	0	0	0	1	1	2
	o	1	0	0	0	2	2
10	*br*	0	1	1	3	6	11
	rp	0	0	0	0	3	3
	he	0	0	2	2	2	6
	o	1	0	0	1	13	14
	o	2	0	0	1	2	3
11	*br*	0	0	0	5	40	45
	rp	0	0	0	1	24	25
	he	0	0	0	3	30	33
	e	0	0	0	0	2	2
	o	1	0	0	0	110	110
	o	2	0	0	1	18	19
	o	3	0	0	0	1	1

(cont.)

Table 8.5 (continued).

CORONA HOLE

h	Type	Δ	D_{2h}	C_{2h}	C_{2v}	C_s	Total
12	br	0	2	5	15	195	217
	rp	0	0	3	7	170	180
	he	0	1	7	12	217	237
	e	0	1	2	0	34	37
	o	1	0	0	5	724	729
	o	2	0	0	3	142	145
	o	3	0	0	1	8	9
13	br	0	0	0	22	953	975
	rp	0	0	0	12	1074	1086
	he	0	0	0	25	1393	1418
	e	0	0	0	2	380	382
	o	1	0	0	2	4535	4537
	o	2	0	0	2	977	979
	o	3	0	0	0	72	72
	o	4	0	0	1	0	1
14	br	0	3	23	66	4366	4458
	rp	0	1	15	46	6180	6242
	he	0	1	38	74	8041	8154
	e	0	1	18	9	3272	3300
	o	1	0	0	26	26893	26919
	o	2	0	0	21	6352	6373
	o	3	0	0	3	544	547
	o	4	0	0	0	10	10
15	o	2	0	0	14	39634	39648
	o	3	0	0	3	3971	3974
	o	4	0	0	2	109	111
16	o	3	0	0	17	27512	27529
	o	4	0	0	4	1047	1051
	o	5	0	0	0	3	3
17	o	4	0	0	10	8886	8896
	o	5	0	0	0	78	78
18	o	5	0	0	3	1072	1075
19	o	6	0	0	2	17	19

*Abbreviations:
    br branched catacondensed;
    e essentially disconnected;
    he half essentially disconnected;
    o non-Kekuléan;
  prm primitive;
    rp regular pericondensed.
Data under the stroke (here $h > 14$; $h > 15$ in Tables 8.6-8.11) are incomplete inasmuch as the numbers for the lower Δ values are unknown.

Table 8.6. Enumeration and detailed classification of coronoids with the
phenalene hole.*

CORONA HOLE							
h	Type	Δ	D_{3h}	C_{3h}	C_{2v}	C_s	Total
9	*prm*	0	1	0	0	0	1
10	*br*	0	0	0	0	1	1
	o	1	0	0	1	1	2
11	*br*	0	0	0	3	4	7
	rp	0	0	0	0	2	2
	he	0	0	0	0	3	3
	o	1	0	0	1	10	11
	o	2	0	0	3	1	4
12	*br*	0	0	1	0	28	29
	rp	0	0	0	0	17	17
	he	0	0	0	1	23	24
	e	0	0	0	1	1	2
	o	1	0	0	5	84	89
	o	2	0	0	1	21	22
	o	3	1	1	0	1	3
13	*br*	0	0	0	11	132	143
	rp	0	0	0	4	121	125
	he	0	0	0	0	169	169
	e	0	0	0	2	25	27
	o	1	0	0	7	580	587
	o	2	0	0	12	164	176
	o	3	0	0	2	16	18
14	*br*	0	0	0	0	647	647
	rp	0	0	0	1	767	768
	he	0	0	0	3	1049	1052
	e	0	0	0	6	284	290
	o	1	0	0	28	3684	3712
	o	2	0	0	10	1172	1182
	o	3	0	0	3	146	149
	o	4	0	0	2	1	3
15	*br*	0	1	3	40	2950	2994
	rp	0	0	2	20	4443	4465
	he	0	0	1	1	6081	6083
	e	0	0	0	15	2517	2532
	o	1	0	0	38	22175	22213
	o	2	0	0	62	7619	7681
	o	3	1	4	12	1098	1115
	o	4	0	0	1	32	33

(cont.)

Table 8.6 (continued).

	CORONA HOLE						
h	Type	Δ	D_{3h}	C_{3h}	C_{2v}	C_s	Total
16	o	3	0	0	21	7729	7750
	o	4	0	0	8	339	347
	o	5	0	0	0	1	1
17	o	4	0	0	18	3063	3081
	o	5	0	0	2	23	25
18	o	5	0	0	2	370	372
	o	6	1	0	0	0	1
19	o	6	0	0	0	8	8

*See footnote to Table 8.5.

h=9 h=10

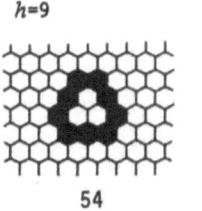

54

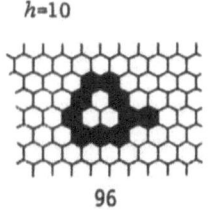

96

h=11

108   123   138   150

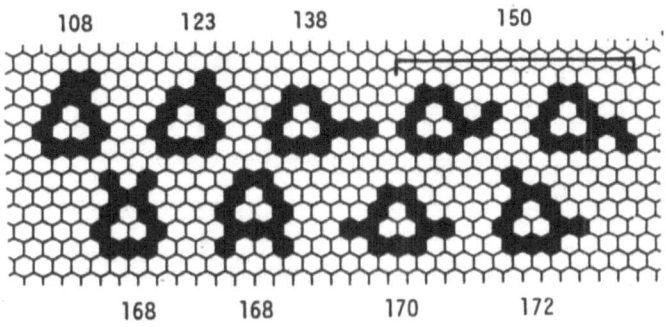

168   168   170   172

(cont.)

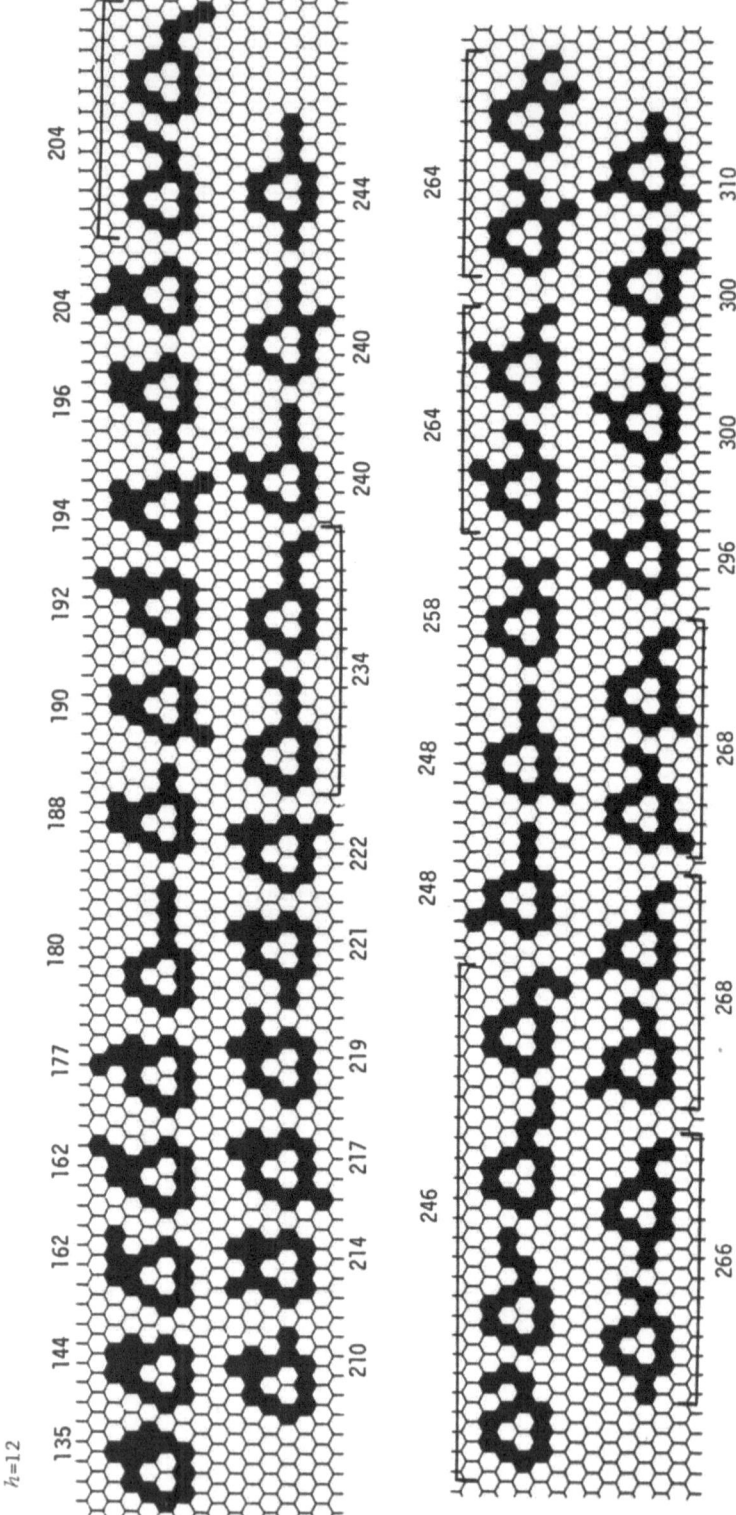

Fig. 8.6. All regular coronoids with the *phenalene* hole for $h \leq 12$. K numbers are given.

8.4.3 *Detailed Classification*

The enumerated coronoids with seven different holes were classified into categories as shown in Tables 5-11. All the basic coronoids with $h \leq 11$ were taken into account. They are primitive coronoids exclusively. To be specific, the different holes are *naphthalene*, *phenalene*, *anthracene*, *pyrene*, *phenanthrene*, *naphthanthrene* and finally *benzophenalene*.

The main principles of the classification have been treated in the preceding sections.

The results of the enumerations and classifications of the coronoids with *naphthalene* hole are shown in Table 5. Most of the forms (but not all of them) in Figs. 7.1 - 7.6 have the *naphthalene* hole.

Table 6 shows the corresponding results for coronoids with *phenalene* hole. We have also included figures of some forms of such systems: the regular coronoids with h = 9, 10, 11 and 12 (Fig. 6); the HED coronoids with h = 11 and 12 (Fig. 7); the

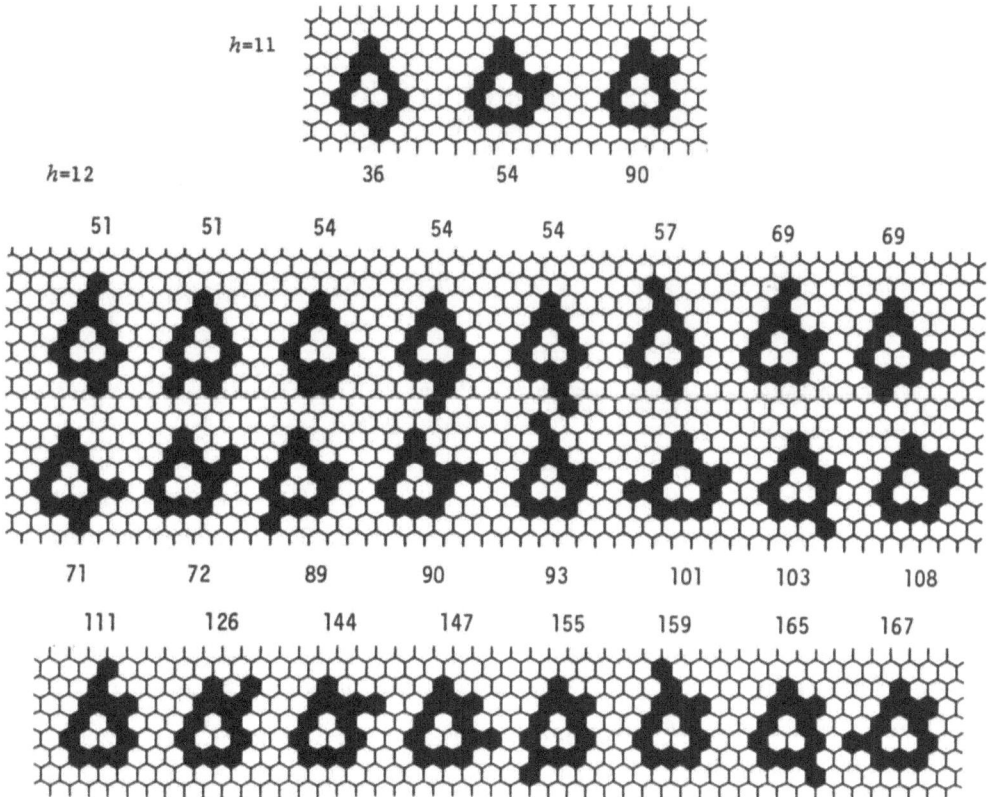

Fig. 8.7. All HED (half essentially disconnected) coronoids with the *phenalene* hole for $h \leq 12$. K numbers are given.

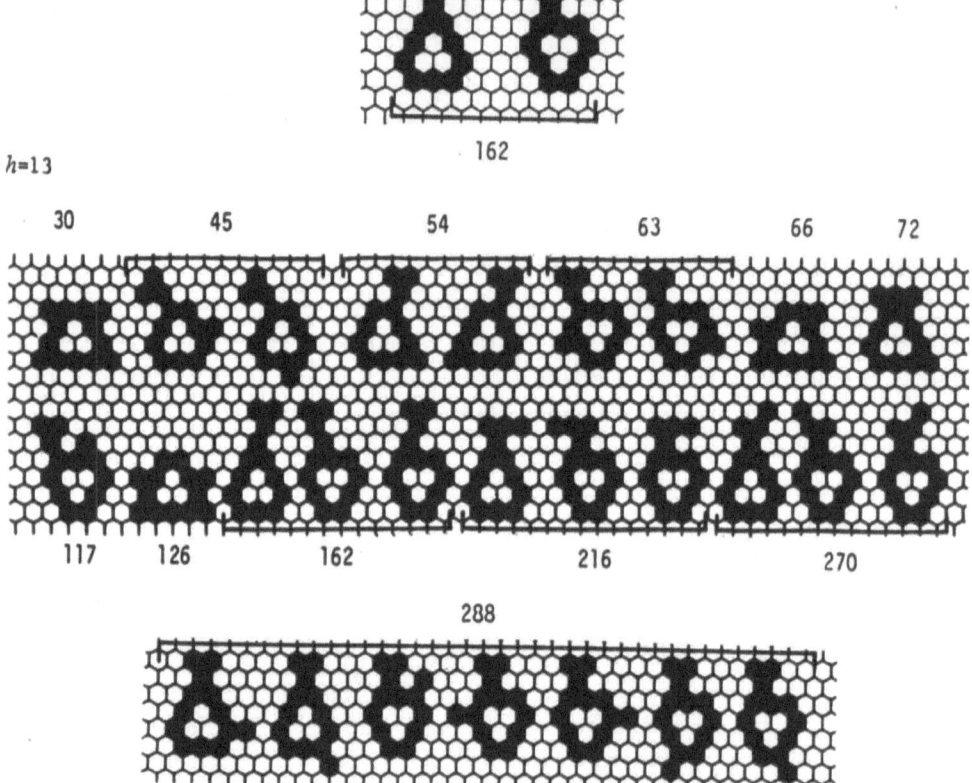

Fig. 8.8. All essentially disconnected coronoids with the *phenalene* hole for $h \leq 13$. K numbers are given.

essentially disconnected coronoids with $h = 12$ and 13 (Fig. 8); the obvious non-Kekuléan coronoids with $\Delta=1$ and $h = 10$, with $\Delta=1$ and $h = 11$, and finally $\Delta=2$, $h = 11$ (Fig. 9). Further forms of obvious non-Kekuléan coronoids with the *phenalene* hole are found among the extremal systems described in Paragraph 5.4.3: the three systems with $h = 12$ and $\Delta=3$ (Fig. 5.4); the three systems with $h = 14$ and $\Delta=4$ (Fig. 5.4); the unique systems with $h = 16$, $\Delta=5$, with $h = 18$, $\Delta=6$, and with $h = 21$, $\Delta=7$ (Fig. 5.5); finally the seven systems with $h = 24$ and $\Delta=8$ (Fig. 5.5). The concealed non-Kekuléan coronoid systems with the *phenalene* hole and $h = 16$ were detected by the method of sifting as described in Section 7.3 (for the *naphthalene* hole). The resulting 21 systems are depicted in Fig. 10.

The remaining tables of this section show the enumeration results for coronoids with the other holes mentioned above, viz.: *anthracene* (Table 7), *pyrene* (Table 8), *phenanthrene* (Table 9), *naphthanthrene* (Table 10) and *benzophenalene* (Table 11).

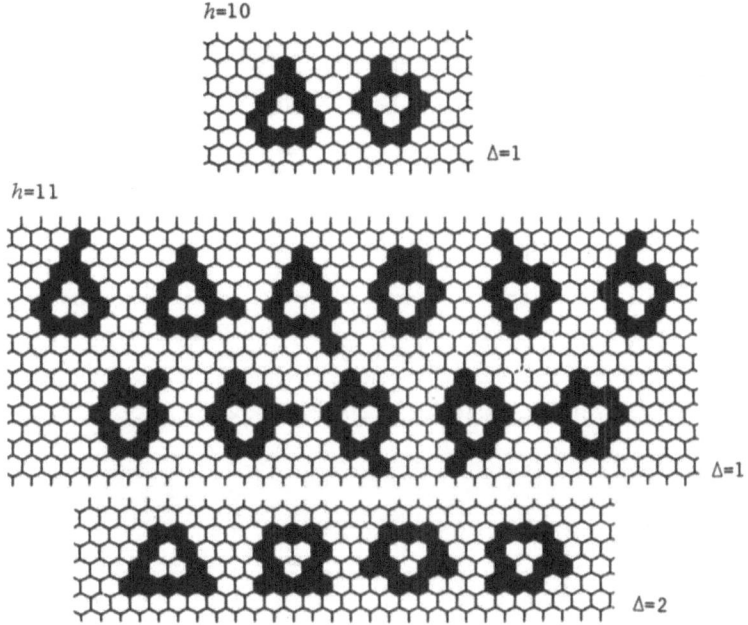

Fig. 8.9. All non-Kekuléan coronoids with the *phenalene* hole for $h \leq 11$.

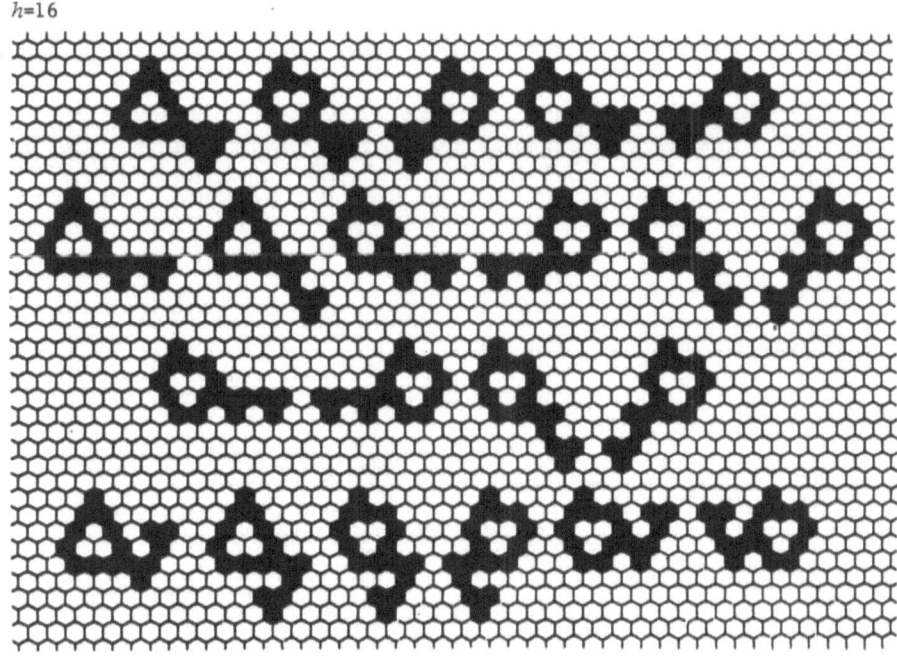

Fig. 8.10. The smallest ($h = 16$) concealed non-Kekuléan coronoids with the *phenalene* hole.

Table 8.7. Enumeration and detailed classification of coronoids with the *anthracene* hole.*

h	Type	Δ	D_{2h}	C_{2h}	C_{2v}	C_s	Total
10	*prm*	0	1	0	0	0	1
11	*br*	0	0	0	1	1	2
	o	1	0	0	1	2	3
12	*br*	0	1	1	3	6	11
	rp	0	0	0	0	3	3
	he	0	1	2	2	4	9
	o	1	0	0	0	18	18
	o	2	0	0	2	3	5
13	*br*	0	0	0	5	40	45
	rp	0	0	0	1	24	25
	he	0	0	0	4	52	56
	e	0	0	0	1	2	3
	o	1	0	0	6	153	159
	o	2	0	0	0	32	32
	o	3	0	0	1	2	3
14	*br*	0	2	5	15	195	217
	rp	0	0	3	8	172	183
	he	0	1	10	19	397	427
	e	0	2	3	1	49	55
	o	1	0	0	2	1077	1079
	o	2	0	0	7	256	263
	o	3	0	0	0	22	22
	o	4	0	0	1	0	1
15	*br*	0	0	0	22	955	977
	rp	0	0	0	12	1102	1114
	he	0	0	0	35	2661	2696
	e	0	0	0	6	586	592
	o	1	0	0	31	6972	7003
	o	2	0	0	4	1819	1823
	o	3	0	0	5	181	186
	o	4	0	0	0	5	5
16	*o*	2	0	0	44	11986	12030
	o	3	0	0	3	1371	1374
	o	4	0	0	3	49	52
17	*o*	3	0	0	26	9702	9728
	o	4	0	0	2	448	450
	o	5	0	0	0	3	3
18	*o*	4	0	0	20	3706	3726
	o	5	0	0	1	48	49
19	*o*	5	0	0	3	580	583
20	*o*	6	0	0	3	17	20

*See footnote to Table 8.5.

Table 8.8. Enumeration and detailed classification of coronoids with the
pyrene hole.*

	CORONA HOLE						
h	Type	Δ	D_{2h}	C_{2h}	C_{2v}	C_s	Total
10	prm	0	1	0	0	0	1
11	br	0	0	0	1	1	2
	o	1	0	0	1	2	3
12	br	0	1	1	3	6	11
	rp	0	0	0	0	3	3
	he	0	1	0	2	4	9
	o	1	0	0	0	18	18
	o	2	0	0	2	3	5
13	br	0	0	0	5	40	45
	rp	0	0	0	1	26	27
	he	0	0	0	4	52	56
	e	0	0	0	1	2	3
	o	1	0	0	5	150	155
	o	2	0	0	0	34	34
	o	3	0	0	2	1	3
14	br	0	2	5	14	197	218
	rp	0	0	3	7	189	199
	he	0	1	10	16	389	416
	e	0	2	3	2	44	51
	o	1	0	0	2	1067	1069
	o	2	0	0	12	261	273
	o	3	0	0	0	21	21
15	br	0	0	0	22	966	988
	rp	0	0	0	15	1220	1235
	he	0	0	0	32	2584	2616
	e	0	0	0	6	541	547
	o	1	0	0	28	6929	6957
	o	2	0	0	4	1866	1870
	o	3	0	0	8	173	181
	o	4	0	0	0	2	2
16	o	2	0	0	51	12281	12332
	o	3	0	0	4	1334	1338
	o	4	0	0	2	31	33
17	o	3	0	0	34	9572	9606
	o	4	0	0	0	339	339
	o	5	0	0	1	0	1
18	o	4	0	0	16	3094	3110
	o	5	0	0	0	17	17
19	o	5	0	0	7	300	307
20	o	6	0	0	0	3	3

*See footnote to Table 8.5.

Table 8.9. Enumeration and detailed classification of coronoids with the *phenanthrene* hole.*

CORONA HOLE					

h	Type	Δ	C_{2v}	C_s	Total
10	prm	0	1	0	1
11	br	0	1	3	4
	rp	0	1	0	1
	o	1	0	4	4
12	br	0	4	19	23
	rp	0	2	8	10
	he	0	4	6	10
	o	1	0	36	36
	o	2	0	5	5
13	br	0	5	111	116
	rp	0	6	74	80
	he	0	8	79	87
	e	0	0	4	4
	o	1	0	297	297
	o	2	0	50	50
	o	3	0	2	2
14	br	0	16	564	580
	rp	0	19	542	561
	he	0	30	633	663
	e	0	2	82	84
	o	1	0	2095	2095
	o	2	0	415	415
	o	3	0	23	23
15	br	0	22	2756	2778
	rp	0	45	3357	3402
	he	0	56	4384	4440
	e	0	3	975	978
	o	1	0	13659	13659
	o	2	0	3004	3004
	o	3	0	217	217
	o	4	0	1	1
16	o	2	0	20315	20315
	o	3	0	1756	1756
	o	4	0	27	27
17	o	3	0	13254	13254
	o	4	0	339	339
18	o	4	0	3445	3445
	o	5	0	8	8
19	o	5	0	231	231

*See footnote to Table 8.5.

Table 8.10. Enumeration and detailed classification of coronoids with the *naphthanthrene* hole.*

	CORONA HOLE	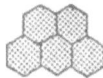			

h	Type	Δ	C_{2v}	C_s	Total
11	prm	0	1	0	1
12	br	0	0	3	3
	o	1	1	5	6
13	br	0	3	15	18
	rp	0	0	6	6
	he	0	0	14	14
	o	1	1	40	41
	o	2	4	9	13
14	br	0	0	85	85
	rp	0	0	53	53
	he	0	1	121	122
	e	0	1	5	6
	o	1	5	350	355
	o	2	1	99	100
	o	3	2	8	10
15	br	0	12	408	420
	rp	0	5	388	393
	he	0	0	917	917
	e	0	3	105	108
	o	1	9	2507	2516
	o	2	15	804	819
	o	3	1	102	103
	o	4	2	1	3
16	o	2	10	5813	5823
	o	3	11	883	894
	o	4	1	33	34
17	o	3	11	6717	6728
	o	4	10	357	367
	o	5	0	2	2
18	o	4	5	3244	3249
	o	5	1	40	41
19	o	5	3	538	541
	o	6	1	0	1
20	o	6	0	19	19

*See footnote to Table 8.5.

Table 8.11. Enumeration and detailed classification of coronoids with the *benzophenalene* hole.*

	CORONA HOLE		

h	Type	Δ	C_s = Total
11	*prm*	0	1
12	*br*	0	7
	rp	0	1
	o	1	9
13	*br*	0	42
	rp	0	18
	he	0	18
	o	1	83
	o	2	15
14	*br*	0	227
	rp	0	154
	he	0	186
	e	0	9
	o	1	701
	o	2	162
	o	3	10
15	*br*	0	1144
	rp	0	1108
	he	0	1461
	e	0	192
	o	1	5029
	o	2	1394
	o	3	128
	o	4	2
16	*o*	2	10283
	o	3	1218
	o	4	33
17	*o*	3	9767
	o	4	399
	o	5	1
18	*o*	4	3864
	o	5	32
19	*o*	5	500
20	*o*	6	8

*See footnote to Table 8.5.

Chapter 9

ENUMERATION: NUMBERS OF INTERNAL VERTICES

9.1 INTRODUCTION

In the preceding chapters the classification of the coronoids into catacondensed and pericondensed systems is taken into account. This means a classification according to $n_i = 0$ and $n_i > 0$, respectively, where n_i is the number of internal vertices. A more detailed classification of the pericondensed coronoid systems according to the values of n_i is also possible. This kind of classification, which is the subject of the present chapter, is of a particular chemical interest because it is virtually an enumeration of the chemical isomers. Some enumerations of this kind for coronoid systems, beyond the catacondensed coronoids, have been reported (Cyvin SJ and Brunvoll 1989; Dias 1990; Cyvin SJ, Brunvoll and Cyvin 1990b).

9.2 CORONOID CHEMICAL ISOMERS

A coronoid system corresponds to a coronoid hydrocarbon (chemically known or unknown) with the formula:

$$C_n H_{n_2}$$

Here n, the number of vertices, corresponds to the number of carbon atoms, while n_2, the number of vertices of degree two, corresponds to the number of hydrogens.

In an enumeration of *coronoid* (chemical) *isomers* the coronoid systems are counted which are compatible with a given formula. The coefficients of this formula (n, n_2) are related to other pairs of invariants, e.g. (h, n_i):

$$n = 4h - n_i, \qquad n_2 = 2h - n_i \qquad (9.1)$$

cf. eqns. (3.1) and (3.3). Therefore the classification of a set of coronoids with a given h according to the number of internal vertices (n_i) can readily be translated to the numbers of coronoid isomers.

Similarly an enumeration with h as the leading parameter and a classification according to n_b, the number of boundary vertices or the combined (outer and inner) perimeter lengths, as virtually an enumeration of coronoid isomers. One has the relations:

Table 9.1. Numbers of coronoids with the *naphthalene* hole ($h^o = 2$, $n_i^o = 0$), classified according to the numbers of internal vertices (n_i) and *neo*.*

h	n_i	Δ=0		o (non-Kekulén)				Total
		n	e	Δ=1	2	3	4	
8	0	1						1
9	0	2						2
	1	0		2				2
10	0	11		0				11
	1	0		13				13
	2	9		0	3			12
	3	0		1	0			1
11	0	45		0	0			45
	1	0		76	0			76
	2	53	2	0	17			72
	3	0	0	34	0	1		35
	4	5	0	0	2	0		7
12	0	217	0	0	0	0		217
	1	0	0	412	0	0		412
	2	322	30	0	110	0		462
	3	0	0	282	0	7		289
	4	93	7	0	35	0		135
	5	0	0	35	0	2		37
	6	2	0	0	0	0		2
13	0	975	0	0	0	0		975
	1	0	0	2109	0	0		2109
	2	1714	270	0	641	0		2625
	3	0	0	1947	0	56		2003
	4	689	104	0	292	0		1085
	5	0	0	464	0	16		480
	6	101	8	0	46	0	1	156
	7	0	0	17	0	0	0	17
14	0	4458	0	0	0	0	0	4458
	1	0	0	10469	0	0	0	10469
	2	8724	1952	0	3598	0	0	14274
	3	0	0	12030	0	396	0	12426
	4	4491	1205	0	2217	0	7	7920
	5	0	0	3871	0	127	0	3998
	6	1103	140	0	525	0	3	1771
	7	0	0	545	0	24	0	569
	8	78	3	0	33	0	0	114
	9	0	0	4	0	0	0	4

*Abbreviations: e essentially disconnected; n normal; o non-Kekulén (obvious non-Kekuléans with the Δ values included).

$$n = 2h + \frac{1}{2}\, n_b\, , \qquad\qquad n_2 = \frac{1}{2}\, n_b \qquad\qquad\qquad (9.2)$$

The invariant n_b is given in terms of the original pair of invariants $(h,\ n_i)$ by

$$n_b = 4h - 2n_i \qquad\qquad\qquad (9.3)$$

cf. eqn. (3.5).

9.3 COMPUTATIONAL RESULTS

In a computation for coronoids with definite holes (cf. Section 8.4) a classification according to the numbers of internal vertices (n_i) was executed. Only complete data are included herein; for each corona hole and h value they cover all the
possible n_i values.

Tables 1-7 correspond to the data in Tables 8.5 - 8.11 above the strokes and
pertain to the holes: *naphthalene* (Table 1), *phenalene* (Table 2), *anthracene* (Table
3), *pyrene* (Table 4), *phenanthrene* (Table 5), *naphthanthrene* (Table 6) and *benzophenalene* (Table 7).

A corona hole is characterized by the invariants h^o and $n_i{}^o$ as explained in
Paragraph 3.2.2. In Table 8 each pair of these invariants $(h^o,\ n_i{}^o)$ represents in
most cases more than one benzenoid, as also is the case for $h^o = 3$, $n_i{}^o = 0$ (Tables
3 and 5). The numbers for the basic coronoids with $h = 15$ are listed in Table 9.
Notice that the entry $h^o = 6$, $n_i{}^o = 1$ appears in both Tables 8 and 9. Here Table 8
gives the total numbers, which for $h = 15$ include those for the basic coronoids listed in Table 9.

Altogether Tables 1-8 contain the pertinent data for all coronoids with $h \leq 14$.
If Table 9 is added, one obtains data for the coronoids with $h = 15$ and $h^o > 2$, i.e.
for all the coronoids with $h = 15$ except those with the *naphthalene* hole. So far
these 321409 systems (cf. Table 8.3 or 8.4) have not been generated specifically,
which would have allowed for a classification according to n_i.

Table 9.2. Numbers of coronoids with the *phenalene* hole ($h^o = 3$, $n_i^o = 1$), classified according to the numbers of internal vertices (n_i) and *neo*.*

h	n_i	$\Delta=0$		o (non-Kekuléan)				Total
		n	e	$\Delta=1$	2	3	4	
9	0	1						1
10	0	1						1
	1	0		2				2
11	0	7		0				7
	1	0		10				10
	2	5		0	4			9
	3	0		1	0			1
12	0	29		0	0			29
	1	0		60	0			60
	2	37	2	0	19			58
	3	0	0	29	0	3		32
	4	4	0	0	3	0		7
13	0	143	0	0	0	0		143
	1	0	0	320	0	0		320
	2	225	23	0	131	0		379
	3	0	0	234	0	13		247
	4	68	4	0	44	0		116
	5	0	0	33	0	5		38
	6	1	0	0	1	0		2
14	0	647	0	0	0	0		647
	1	0	0	1643	0	0		1643
	2	1225	210	0	740	0		2175
	3	0	0	1633	0	105		1738
	4	518	73	0	374	0		965
	5	0	0	417	0	42		459
	6	77	7	0	68	0	3	155
	7	0	0	19	0	2	0	21
15	0	2994	0	0	0	0	0	2994
	1	0	0	8128	0	0	0	8128
	2	6244	1513	0	4115	0	0	11872
	3	0	0	10110	0	716	0	10826
	4	3379	908	0	2764	0	13	7064
	5	0	0	3458	0	330	0	3788
	6	857	109	0	744	0	17	1727
	7	0	0	512	0	68	0	580
	8	68	2	0	58	0	3	131
	9	0	0	5	0	1	0	6

*See footnote to Table 9.1.

Table 9.3. Numbers of coronoids with the *anthracene* hole ($h^o = 3$, $n_i^o = 0$), classified according to the numbers of internal vertices (n_i) and *neo*.*

h	n_i	Δ=0 n	Δ=0 e	o (non-Kekuléan) Δ=1	2	3	4	Total
10	0	1						1
11	0	2						2
	1	0		3				3
12	0	11		0				11
	1	0		17				17
	2	12		0	5			17
	3	0		1	0			1
13	0	45		0	0			45
	1	0		103	0			103
	2	73	3	0	29			105
	3	0	0	55	0	3		58
	4	8	0	0	3	0		11
	5	0	0	1	0	0		1
14	0	217	0	0	0	0		217
	1	0	0	557	0	0		557
	2	455	39	0	189	0		683
	3	0	0	461	0	19		480
	4	148	16	0	72	0	1	237
	5	0	0	61	0	3	0	64
	6	7	0	0	2	0	0	9
15	0	977	0	0	0	0	0	977
	1	0	0	2872	0	0	0	2872
	2	2471	362	0	1099	0	0	3932
	3	0	0	3218	0	139	0	3357
	4	1147	211	0	622	0	4	1984
	5	0	0	859	0	44	0	903
	6	190	19	0	101	0	1	311
	7	0	0	54	0	3	0	57
	8	2	0	0	1	0	0	3

*See footnote to Table 9.1.

Table 9.4. Numbers of coronoids with the *pyrene* hole ($h^o = 4$, $n_i^o = 2$), classified according to the numbers of internal vertices (n_i) and *neo*.*

h	n_i	Δ=0		o (non-Kekuléan)				Total
		n	e	Δ=1	2	3	4	
10	0	1						1
11	0	2						2
	1	0		3				3
12	0	11		0				11
	1	0		17				17
	2	12		0	5			17
	3	0		1	0			1
13	0	45		0	0			45
	1	0		102	0			102
	2	74	3	0	30			107
	3	0	0	53	0	3		56
	4	9	0	0	4	0		13
14	0	218	0	0	0	0		218
	1	0	0	548	0	0		548
	2	463	39	0	201	0		703
	3	0	0	450	0	16		466
	4	148	12	0	71	0		231
	5	0	0	71	0	5		76
	6	4	0	0	1	0		5
15	0	988	0	0	0	0		988
	1	0	0	2814	0	0		2814
	2	2507	356	0	1161	0		4024
	3	0	0	3208	0	131		3339
	4	1150	168	0	594	0		1912
	5	0	0	890	0	48		938
	6	193	23	0	115	0	2	333
	7	0	0	45	0	2	0	47
	8	1	0	0	0	0	0	1

*See footnote to Table 9.1.

Table 9.5. Numbers of coronoids with the *phenanthrene* hole ($h^{\circ} = 3$, $n_i^{\circ} = 0$; see also Table 9.3), classified according to the numbers of internal vertices (n_i) and neo.*

h	n_i	$\Delta=0$		o (non-Kekuléan)				Total
		n	e	$\Delta=1$	2	3	4	
10	0	1						1
11	0	4						4
	1	0		4				4
	2	1		0				1
12	0	23		0				23
	1	0		30				30
	2	20		0	5			25
	3	0		6	0			6
13	0	116		0	0			116
	1	0		191	0			191
	2	137	4	0	39			180
	3	0	0	104	0	2		106
	4	30	0	0	11	0		41
	5	0	0	2	0	0		2
14	0	580	0	0	0	0		580
	1	0	0	1091	0	0		1091
	2	864	68	0	270	0		1202
	3	0	0	838	0	16		854
	4	339	16	0	137	0		492
	5	0	0	166	0	7		173
	6	21	0	0	8	0		29
15	0	2778	0	0	0	0		2778
	1	0	0	5822	0	0		5822
	2	4834	657	0	1684	0		7175
	3	0	0	5781	0	137		5918
	4	2518	293	0	1091	0		3902
	5	0	0	1886	0	70		1956
	6	482	28	0	228	0	1	739
	7	0	0	170	0	10	0	180
	8	8	0	0	1	0	0	9

*See footnote to Table 9.1.

Table 9.6. Numbers of coronoids with the *naphthanthrene* hole ($h^o = 5$, $n_i = 3$), classified according to the numbers of internal vertices (n_i) and *neo*.*

h	n_i	Δ=0		o (non-Kekuléan)				Total
		n	e	Δ=1	2	3	4	
11	0	1						1
12	0	3						3
	1	0		6				6
13	0	18		0				18
	1	0		38				38
	2	20		0	13			33
	3	0		3	0			3
14	0	85		0	0			85
	1	0		229	0			229
	2	156	6	0	87			249
	3	0	0	125	0	10		135
	4	19	0	0	13	0		32
	5	0	0	1	0	0		1
15	0	420	0	0	0	0		420
	1	0	0	1241	0	0		1241
	2	977	87	0	586	0		1650
	3	0	0	1099	0	81		1180
	4	324	21	0	227	0	3	575
	5	0	0	176	0	22	0	198
	6	9	0	0	6	0	0	15

*
See footnote to Table 9.1.

Table 9.7. Numbers of coronoids with the *benzophenalene* hole ($h^o = 4$, $n_i^o = 1$), classified according to the numbers of internal vertices (n_i) and *neo*.*

h	n_i	$\Delta=0$		o (non-Kekuléan)				Total
		n	e	$\Delta=1$	2	3	4	
11	0	1						1
12	0	7						7
	1	0		9				9
	2	1		0				1
13	0	42		0				42
	1	0		69				69
	2	36		0	15			51
	3	0		14	0			14
14	0	227		0	0			227
	1	0		444	0			444
	2	282	9	0	127			418
	3	0	0	251	0	10		261
	4	58	0	0	35	0		93
	5	0	0	6	0	0		6
15	0	1144	0	0	0	0		1144
	1	0	0	2553	0	0		2553
	2	1798	156	0	898	0		2852
	3	0	0	2061	0	92		2153
	4	725	36	0	464	0	2	1227
	5	0	0	414	0	36	0	450
	6	46	0	0	32	0	0	78
	7	0	0	1	0	0	0	1

* See footnote to Table 9.1.

Table 9.8. Numbers of coronoids with different holes, classified according to the numbers of internal vertices (n_i) and neo.*

h^o	n_i^o	h	n_i	Δ=0 n	Δ=0 e	o (non-Kekuléan) Δ=1	2	3	4	Total
4	0	12	0	4						4
			1	0		1				1
		13	0	15		0				15
			1	0		21				21
			2	7		0	2			9
		14	0	94		0	0			94
			1	0		159	0			159
			2	118		0	43			161
			3	0		60	0	3		63
			4	5		0	1	0		6
		15	0	490		0	0	0		490
			1	0		1003	0	0		1003
			2	851	21	0	314	0		1186
			3	0	0	815	0	37		852
			4	220	13	0	106	0	1	340
			5	0	0	64	0	2	0	66
			6	2	0	0	0	0	0	2
5	2	12	0	3	0	0	0	0	0	3
		13	0	14	0	0	0	0	0	14
			1	0	0	16	0	0	0	16
			2	3	0	0	0	0	0	3
		14	0	87	0	0	0	0	0	87
			1	0	0	129	0	0	0	129
			2	79	0	0	24	0	0	103
			3	0	0	33	0	0	0	33
			4	2	0	0	0	0	0	2
		15	0	455	0	0	0	0	0	455
			1	0	0	856	0	0	0	856
			2	618	16	0	200	0	0	834
			3	0	0	526	0	12	0	538
			4	156	0	0	67	0	0	223
			5	0	0	22	0	0	0	22
6	4	12	0	3	0	0	0	0	0	3
		13	0	8	0	0	0	0	0	8
			1	0	0	14	0	0	0	14
			2	1	0	0	0	0	0	1
		14	0	48	0	0	0	0	0	48
			1	0	0	95	0	0	0	95
			2	59	0	0	28	0	0	87
			3	0	0	12	0	0	0	12

(cont.)

Table 9.8 (continued).

h^o	n_i^o	h	n_i	Δ=0		o (non-Kekuléan)				Total
				n	e	Δ=1	2	3	4	
(6	4)	15	0	230	0	0	0	0	0	230
			1	0	0	592	0	0	0	592
			2	425	14	0	208	0	0	647
			3	0	0	354	0	24	0	378
			4	74	0	0	40	0	0	114
			5	0	0	5	0	0	0	5
7	6	12	0	1	0	0	0	0	0	1
		13	0	1	0	0	0	0	0	1
			1	0	0	1	0	0	0	1
		14	0	5	0	0	0	0	0	5
			1	0	0	7	0	0	0	7
			2	6	0	0	3	0	0	9
			3	0	0	1	0	0	0	1
		15	0	17	0	0	0	0	0	17
			1	0	0	42	0	0	0	42
			2	35	1	0	16	0	0	52
			3	0	0	25	0	2	0	27
			4	5	0	0	3	0	0	8
5	1	13	0	4	0	0	0	0	0	4
			1	0	0	2	0	0	0	2
		14	0	27	0	0	0	0	0	27
			1	0	0	43	0	0	0	43
			2	17	0	0	8	0	0	25
		15	0	174	0	0	0	0	0	174
			1	0	0	330	0	0	0	330
			2	237	0	0	115	0	0	352
			3	0	0	169	0	14	0	183
			4	12	0	0	3	0	0	15
5	0	13	0	1	0	0	0	0	0	1
		14	0	12	0	0	0	0	0	12
			1	0	0	8	0	0	0	8
			2	1	0	0	0	0	0	1
		15	0	62	0	0	0	0	0	62
			1	0	0	112	0	0	0	112
			2	54	0	0	19	0	0	73
			3	0	0	13	0	0	0	13

(cont.)

Table 9.8 (continued).

h^o	n_i^o	h	n_i	$\Delta=0$		o (non-Kekuléan)				Total
				n	e	$\Delta=1$	2	3	4	
6	3	13	0	4	0	0	0	0	0	4
		14	0	26	0	0	0	0	0	26
			1	0	0	36	0	0	0	36
			2	5	0	0	0	0	0	5
		15	0	164	0	0	0	0	0	164
			1	0	0	290	0	0	0	290
			2	160	0	0	69	0	0	229
			3	0	0	71	0	0	0	71
			4	3	0	0	0	0	0	3
7	5	13	0	3	0	0	0	0	0	3
		14	0	17	0	0	0	0	0	17
			1	0	0	29	0	0	0	29
			2	2	0	0	0	0	0	2
		15	0	102	0	0	0	0	0	102
			1	0	0	220	0	0	0	220
			2	118	0	0	62	0	0	180
			3	0	0	40	0	0	0	40
8	7	13	0	1	0	0	0	0	0	1
		14	0	3	0	0	0	0	0	3
			1	0	0	7	0	0	0	7
		15	0	18	0	0	0	0	0	18
			1	0	0	46	0	0	0	46
			2	26	0	0	18	0	0	44
			3	0	0	4	0	0	0	4
6	2	14	0	9	0	0	0	0	0	9
			1	0	0	5	0	0	0	5
		15	0	54	0	0	0	0	0	54
			1	0	0	90	0	0	0	90
			2	46	0	0	21	0	0	67
			3	0	0	3	0	0	0	3
6	1	14	0	2	0	0	0	0	0	2
		15	0	26	0	0	0	0	0	26
			1	0	0	32	0	0	0	32
			2	2	0	0	0	0	0	2

(cont.)

Table 9.8 (continued).

h^o	n_i^o	h	n_i	$\Delta=0$		o (non-Kekuléan)				Total
				n	e	$\Delta=1$	2	3	4	
7	4	14	0	9	0	0	0	0	0	9
			1	0	0	1	0	0	0	1
		15	0	61	0	0	0	0	0	61
			1	0	0	87	0	0	0	87
			2	23	0	0	5	0	0	28
8	6	14	0	9	0	0	0	0	0	9
		15	0	44	0	0	0	0	0	44
			1	0	0	66	0	0	0	66
			2	9	0	0	0	0	0	9
9	8	14	0	4	0	0	0	0	0	4
		15	0	16	0	0	0	0	0	16
			1	0	0	29	0	0	0	29
			2	2	0	0	0	0	0	2
10	10	14	0	1	0	0	0	0	0	1
		15	0	2	0	0	0	0	0	2
			1	0	0	4	0	0	0	4

*See footnote to Table 9.1.

Table 9.9. Numbers of basic coronoids with $h = 15$, classified according to the numbers of internal vertices (n_i) and Δ values.

h^o	n_i^o	$h=15$ n_i	Δ=0 normal	non-Kekuléan Δ=1	Total
6	1	0	12		12
		1	0	10	10
6	0	0	3	0	3
		1	0	1	1
7	3	0	14	0	14
		1	0	11	11
7	2	0	8	0	8
8	5	0	16	0	16
		1	0	5	5
8	4	0	2	0	2
9	7	0	14	0	14
		1	0	1	1
10	9	0	9	0	9
11	11	0	2	0	2
12	13	0	1	0	1

9.4 CORONA HOLES AND THEIR INVARIANTS

9.4.1 *Numbers of Corona Holes*

Let a corona hole be characterized by the invariants h^o and n_i^o as in Paragraph 3.2.2 and the above tables (see also Section 9.3). An enumeration of non-isomorphic corona holes with a given pair of these invariants (h^o, n_i^o) is equivalent to an enumeration of benzenoids with a given number of hexagons and a given number of internal vertices. This is virtually an enumeration of benzenoid (chemical) isomers (Gutman and Cyvin 1989b; Brunvoll and Cyvin 1990). The results of these enumerations for the types of corona holes which pertain to Tables 1-9 are collected in Table 10.

The numbers of Table 10 can be extracted from different sources (Elk 1980; Knop, Szymanski, Jeričević and Trinajstić 1983; Stojmenović et al. 1986). All of them are found in the review of Brunvoll and Cyvin (1990).

9.4.2 *Extremal Values Associated with Corona Holes*

Let H and N_i denote the numbers of hexagons and numbers of internal vertices, respectively, of the benzenoid obtained when the corona hole of a coronoid is filled with hexagons. The same notation is applied in Paragraph 3.2.3. Then one has

$$H = h + h^\circ \tag{9.4}$$

and

$$N_i = n_i + 4h^\circ - n_i^\circ + 2 \tag{9.5}$$

when the other symbols have the same meaning as above.

For the number of internal vertices (n_i) of a coronoid eqn. (3.13) gives the

Table 9.10. Numbers of corona holes.

h°	n_i°	Number	h°	n_i°	Number
2	0	1[a]	7	2	68[b]
				3	25[b]
3	0	2[a]		4	10[b]
	1	1[a]		5	3[b]
				6	1[b]
4	0	5[a]			
	1	1[a]	8	4	67[b]
	2	1[a]		5	21[b]
				6	9[b]
5	0	12[a]		7	1[b]
	1	6[a]			
	2	3[a]	9	7	15[b]
	3	1[a]		8	4[b]
6	0	36[b]	10	9	9[b]
	1	24[b]		10	1[b]
	2	14[b]			
	3	4[b]	11	11	2[c]
	4	3[b]			
			12	13	1[c]

[a] Elk SB (1980). Match 8: 121

[b] Knop JV, Szymanski K, Jeričević Ž, Trinajstić N (1983). J Comput Chem 4: 23

[c] Stojmenović I, Tošić R, Doroslovački R (1986). Proceedings of the Sixth Yugoslav Seminar on Graph Theory, Dubrovnik 1985, Novi Sad: 189

absolute maximum, $(n_i)_{max}$, which is realized for the *naphthalene* hole ($h^o = 2$). We shall deduce a generalization for corona holes with h^o hexagons. On introducing the maximum value of N_i according to eqn. (3.12) and the maximum value of n_i^o according to (3.17) one obtains

$$0 \le n_i \le 2h - \{\sqrt{12(h + h^o) - 3}\} - \{\sqrt{12h^o - 3}\} \tag{9.6}$$

Here the upper bound is always realized.

The right-hand side of (6) is the absolute maximum of n_i under the given conditions, in which n_i^o has the maximum value. A more flexible formula with unspecified n_i^o reads

$$0 \le n_i \le n_i^o + 2(h - h^o) - 1 - \{\sqrt{12(h + h^o) - 3}\} \tag{9.7}$$

In this equation, however, the upper bound is not always realized.

In Tables 1-8 most of the upper bounds according to eqn. (7) - for given h^o, n_i^o and h values - are found to be realized. The exceptions are specified below. For $h^o = 5$, $n_i^o = 1$ and $h = 14$ the formula upper bound is 3, while $n_i = 2$ is the true maximum; for $h^o = 6$, $n_i^o = 3$ and $h = 13$ the formula upper bound is 1, while $n_i = 0$ is the only realized value.

The last question, which is taken up here, is: Which kind of holes are compatible with a coronoid having h hexagons? We can at least give the answer with regard to the number of hexagons of the hole, viz. h^o. The maximum of h^o is realized for one or more primitive coronoids. The value of this maximum is given by eqn. (4.16). It is clear that holes with all sizes within the range

$$2 \le h^o \le \left[\frac{h^2}{12} - \frac{h}{2} + 1\right] \tag{9.8}$$

are possible in the considered case.

In this connection we give the last equation, which also pertains to primitive coronoids;

$$h = 2h^o - n_i^o + 4 \tag{9.9}$$

BIBLIOGRAPHY

1 Agranat I, Hess Jr BA, Schaad LJ (1980) Aromaticity of Non-Alternant Annulenoannulenes and of Corannulenes. Pure & Appl Chem 52: 1399

2 Aihara J (1976) On the Number of Aromatic Sextets in a Benzenoid Hydrocarbon. Bull Chem Soc Japan 49: 1429

3 Balaban AT (1969) Chemical Graphs - VII - Proposed Nomenclature of Branched Cata-Condensed Benzenoid Polycyclic Hydrocarbons. Tetrahedron 25: 2949

4 Balaban AT (1971) Chemical Graphs - XII - Configurations of Annulenes. Tetrahedron 27: 6115

5 Balaban AT (1982) Challenging Problems Involving Benzenoid Polycyclics and Related Systems. Pure & Appl Chem 54: 1075

6 Balaban AT (1985) Application of Graph Theory in Chemistry. J Chem Inf Comput Sci 25: 334

7 Balaban AT (1988) Chemical Graphs - Part 49 - Open Problems in the Area of Condensed Polycyclic Benzenoids - Topological Stereoisomers of Coronoids and Congeners. Rev Roumaine Chim 33: 699

8 Balaban AT (1989) Carbon and Its Nets [in] Symmetry 2 Unifying Human Understanding (Hargittai I, Edit). Pergamon Press, Oxford; Computers Math Applic 17: 397

9 Balaban AT, Brunvoll J, Cioslowski J, Cyvin BN, Cyvin SJ, Gutman I, He WC, He WJ, Knop JV, Kovačević M, Müller WR, Szymanski K, Tošić R, Trinajstic N (1987) Enumeration of Benzenoid and Coronoid Hydrocarbons. Z Naturforsch 42a: 863

10 Balaban AT, Brunvoll J, Cyvin SJ (1990) Chemical Graphs - Part 54 - Enumeration of Unbranched Catacondensed Polyhexes with Equidistant Linearly Condensed Segments. Rev Roumaine Chim: in press

11 Balaban AT, Harary F (1968) Chemical Graphs - V - Enumeration and Proposed Nomenclature of Benzenoid Cata-Condensed Polycyclic Aromatic Hydrocarbons. Tetrahedron 24: 2505

12 Balasubramanian K (1987) Computational Graph Theory [in] Graph Theory and Topology in Chemistry (King RB, Rouvray DH, Edit). Elsevier, Amsterdam; Studies in Physical and Theoretical Chemistry 51: 514

13 Balasubramanian K (1989) Reduced Cycle Indices and Their Applications in Enumeration of NMR Signals and Equivalence Classes. J Math Chem 3: 227

14 Balasubramanian K, Kaufman JJ, Koski WS, Balaban AT (1980) Graph Theo-
    retical Characterization and Computer Generation of Certain Carcino-
    genic Benzenoid Hydrocarbons and Identification of Bay Regions.
    J Comput Chem 1: 149

15 Barth WE, Lawton RG (1966) Dibenzo[ghi,mno]Fluoranthene. J Am Chem
    Soc 88: 380

16 Barth WE, Lawton RG (1971) The Synthesis of Corannulene. J Am Chem
    Soc 93: 1730

17 Bergan JL, Cyvin BN, Cyvin SJ (1987) The Fibonacci Numbers, and
    Kekulé Structures of Some Corona-Condensed Benzenoids (Corannulenes).

18 Bonchev D, Balaban AT (1981) Topological Centric Coding and Nomencla-
    ture of Polycyclic Hydrocarbons - 1 - Condensed Benzenoid Systems
    (Polyhexes, Fusenes). J Chem Inf Comput Sci 21: 223

19 Brendsdal E, Cyvin BN, Brunvoll J, Cyvin SJ (1988) Condensed Aromatics
    - Part XXIV - Cyclo[d.e.d.e.e.d.e.d.e.e]Decakisbenzene. Spectrochim
    Acta 44A: 981

20 Brunvoll J, Cyvin BN, Cyvin SJ (1987a) Enumeration and Classification
    of Coronoid Hydrocarbons. J Chem Inf Comput Sci 27: 14

21 Brunvoll J, Cyvin BN, Cyvin SJ (1987b) Enumeration and Classification
    of Benzenoid Hydrocarbons - 2 - Symmetry and Regular Hexagonal Benze-
    noids. J Chem Inf Comput Sci 27: 171

22 Brunvoll J, Cyvin BN, Cyvin SJ, Gutman I (1988a) Essentially Disconnec-
    ted Benzenoids - Enumeration and Classification of Benzenoid Hydrocar-
    bons - IX. Match 23: 209

23 Brunvoll J, Cyvin BN, Cyvin SJ, Gutman I (1988b) Some Benzenoid Hydro-
    carbons with Extremal Properties. Z Naturforsch 43a: 889

24 Brunvoll J, Cyvin BN, Cyvin SJ, Gutman I, Tošić R, Kovačević M (1989)
    Enumeration and Classification of Coronoid Hydrocarbons - Part V -
    Primitive Coronoids. J Mol Struct (Theochem) 184: 165

25 Brunvoll J, Cyvin BN, Cyvin SJ, Knop JV, Müller WR, Szymanski K,
    Trinajstic N (1990) Enumeration and Classification of Coronoid Hydro-
    carbons - Note to a Note. J Mol Struct (Theochem) 207 131

26 Brunvoll J, Cyvin SJ (1990) What do We Know about the Numbers of
    Benzenoid Isomers? Z Naturforsch 45a: 69

27 Brunvoll J, Cyvin SJ, Cyvin BN (1987) Enumeration and Classification
    of Benzenoid Hydrocarbons. J Comput Chem 8: 189

28 Brunvoll J, Cyvin SJ, Cyvin BN, Gutman I (1989) Essentially Disconnec-
    ted Benzenoids - Distribution of K, the Number of Kekulé Structures,
    in Benzenoid Hydrocarbons - VIII. Match 24: 51

29 Ciosłowski J (1985) Additive Nodal Increments for Approximate Calcula-
    tion of the Total (pi)-Electron Energy of Benzenoid Hydrocarbons.
    Theor Chim Acta 68: 315

30 Cioslowski J (1987) Computer Enumeration of Polyhexes Using the Com-
    pact Naming Approach. J Comput Chem 8: 906

31 Ciosłowski J, Turek AM (1985) An Algorithm to Generate the Compact Name of Benzenoid Hydrocarbons. Computers & Chemistry 9: 247

32 Clar E (1964) Polycyclic Hydrocarbons - Vols I, II. Academic Press, London

33 Clar E (1972) The Aromatic Sextet. Wiley, London

34 Clar E, Robertson JM, Schlögl R, Schmidt W (1981) Photoelectron Spectra of Polynuclear Aromatics - 6 - Application to Structural Elucidation - "Circumanthracene". J Am Chem Soc 103: 1320

35 Cram DJ, Dewhirst KC (1959) Macro Rings - XIX - Olefinic Paracyclophanes. J Am Chem Soc 81: 5963

36 Cresp TM, Sondheimer F (1975) 4,9,16,21-Tetramethyl-5,7,17,19-tetrakisdehydro[14]annuleno[14]annulene, a Macrocyclic Analog of Naphthalene. J Am Chem Soc 97: 4412

37 Cresp TM, Sondheimer F (1977) The Synthesis of Derivatives of [14]Annuleno[14]annulene, [14]Annuleno[16]annulene, and [14]Annuleno[18]annulene, Bicyclic Compounds Consisting of Two Ortho Fused Macrocyclic Systems. J Am Chem Soc 99: 194

38 Cvetković D, Gutman I, Trinajstić N (1974) Graph Theory and Molecular Orbitals - VII - The Role of Resonance Structures. J Chem Phys 61: 2700

39 Cyvin BN, Brunvoll J, Cyvin SJ, Gutman I (1986) Distribution of K, the Number of Kekulé Structures in Benzenoid Hydrocarbons - Part III - Kekulé Structure Statistics. Match 21: 301

40 Cyvin SJ (1988) The Number of Kekulé Structures for Primitive Coronoids (Cycloarenes). Chem Phys Letters 147: 384

41 Cyvin SJ (1989) Enumeration of Kekulé Structures for Primitive Coronoid Hydrocarbons - "Waffles". Monatsh Chem 120: 243

42 Cyvin SJ, Bergan JL, Cyvin BN (1987) Benzenoids and Coronoids with Hexagonal Symetry ("Snowflakes"). Acta Chim Hung 124: 691

43 Cyvin SJ, Brunvoll J (1989) Chemical Isomers of Coronoid Hydrocarbons. Chem Phys Letters 164: 635

44 Cyvin SJ, Brunvoll J, Cyvin BN (1988) Molecular Vibrations of Cyclo-[d.e.e.d.e.e.d.e.e]Nonakisbenzene and the Topology of Primitive Coronoids with Trigonal Symmetry. Acta Chem Scand A42: 434

45 Cyvin SJ, Brunvoll J, Cyvin BN (1989a) Topological Aspects of Benzenoids and Coronoids, Including "Snowflakes" and "Laceflowers" [in] Symmetry 2 Unifying Human Understanding (Hargittai I, Edit). Pergamon Press, Oxford; Computers Math Applic 17: 355

46 Cyvin SJ, Brunvoll J, Cyvin BN (1989b) Distribution of K, the Number of Kekulé Structures, in Benzenoid Hydrocarbons - Normal Benzenoids with K to 110. J Chem Inf Comput Sci 29: 79

47 Cyvin SJ, Brunvoll J, Cyvin BN (1989c) Search for Concealed Non-Kekuléan Benzenoids and Coronoids. J Chem Inf Comput Sci 29: 236

48 Cyvin SJ, Brunvoll J, Cyvin BN (1989d) Kekulé Structure Counts of Special Coronoid Hydrocarbons - Hollow Hexagons [in] MATH/CHEM/COMP 1988 (Graovac A, Ed), Elsevier, Amsterdam; Studies in Physical and Theoretical Chemistry 63: 127

49 Cyvin SJ, Brunvoll J, Cyvin BN (1990a) The Hunt for Concealed Non-Kekuléan Polyhexes. J Math Chem: in press

50 Cyvin SJ, Brunvoll J, Cyvin BN (1990b) A Periodic Table for All-Benzenoid Hydrocarbons, and Enumerations of Some Polyhex Isomers. J Math Chem: in press

51 Cyvin SJ, Brunvoll J, Cyvin BN (1990c) Kekulé Structure Counts in Coronoid Hydrocarbons - A General Solution. Struct Chem: in press

52 Cyvin SJ, Brunvoll J, Cyvin BN, Bergan JL, Brendsdal E (1990) Enumeration and Classification of Coronoid Hydrocarbons - Hollow Hexagons. Struct Chem: in press

53 Cyvin SJ, Brunvoll J, Cyvin BN, Brendsdal E (1988) Condensed Aromatics - Part XXIII - Cyclo[d.e.d.e.d.e.d.e.d.e.d.e]Dodekakisbenzene (Kekulene). Spectrochim Acta 44A: 975

54 Cyvin SJ, Brunvoll J, Cyvin BN, Tošić R, Kovačević M (1989) Waffles. J Mol Struct (Theochem) 200: 261

55 Cyvin SJ, Brunvoll J, Gutman I (1990) Annulenes and Primitive Coronoids - Some Topological Properties and Enumerations. Rev Roumaine Chim: in press

56 Cyvin SJ, Cyvin BN, Brunvoll J (1987) Half Essentially Disconnected Coronoid Hydrocarbons. Chem Phys Letters 140: 124

57 Cyvin SJ, Cyvin BN, Brunvoll J (1989) Kekulé Structure Counts and Multiple Coronoid Hydrocarbons. Chem Phys Letters 156: 595

58 Cyvin SJ, Cyvin BN, Brunvoll J, Bergan JL (1987) Coronoid Hydrocarbons with Hexagonal Symmetry. Coll Sci Papers Fac Sci Kragujevac 8: 137

59 Cyvin SJ, Gutman I (1986) Number of Kekulé Structures as a Function of the Number of Hexagons in Benzenoid Hydrocarbons. Z Naturforsch 41a: 1079

60 Cyvin SJ, Gutman I (1988) Kekulé Structures in Benzenoid Hydrocarbons. Springer-Verlag, Berlin; Lecture Notes in Chemistry 46

61 Davidson RA (1981) Spectral Analysis of Graphs by Cyclic Automorphism Subgroups. Theor Chim Acta 58: 193

62 Dias JR (1982) A Periodic Table for Polycyclic Aromatic Hydrocarbons - Isomer Enumeration of Fused Polycyclic Aromatic Hydrocarbons - 1. J Chem Inf Comput Sci 22: 15

63 Dias JR (1983) A Periodic Table for Polycyclic Aromatic Hydrocarbons - Part 3 - Enumeration of all the Polycyclic Conjugated Isomers of Pyrene Having Ring Sizes Ranging from 3 to 9. Match 14: 83

64 Dias JR (1984) A Periodic Table for Polycyclic Aromatic Hydrocarbons - 4 - Isomer Enumeration of Polycyclic Conjugated Hydrocarbons - 2. J Chem Inf Comput Sci 24: 124

65 Dias JR (1988) Characteristic Polynomials and Eigenvalues of Molecular Graphs with a Greater than Twofold Axis of Symmetry. J Mol Struct (Theochem) 165: 125

66 Dias JR (1990) Benzenoid Series Having a Constant Number of Strictly

Peri-Condensed Constant-Isomer Series. J Chem InfComput Sci 30: 251

67 Diederich F, Staab HA (1978) Benzenoid Versus Annulenoid Aromaticity - Synthesis and Properties of Kekulene. Angew Chem Int Ed Engl 17: 372

68 Dopper JE, Wynberg H (1972) Heterocyclic Circulenes - A New Class of Polycyclic Aromatic Hydrocarbons. Tetrahedron Letters: 763

69 Dopper JH, Wynberg H (1975) Synthesis and Properties of Some Hetero-circulenes. J Org Chem 40: 1957

70 Doroslovački R, Tošić R (1984) A Characterization of Hexagonal Systems. Review of Research Fac Sci Univ Novi Sad, Math Ser 14: 201

71 DuVernet RB, Wennerström O, Lawson J, Otsubo T, Boekelheide V (1978) Bridged [18]Annulenes - A Study of the Synthesis and Properties of 12c,12d,12e,12f-Tetrahydrobenzo[g,h,i]perylene and Its Analogues. J Am Chem Soc 100: 2457

72 Džonova-Jerman-Blažič B, Trinajstić N (1982) Computer-Aided Enumeration and Generation of the Kekulé Structures in Conjugated Hydrocarbons. Computers & Chemistry 6: 121

73 Ege G, Fischer H (1967) Zur Konjugation in makrocyclischen Bindungssystemen - VI - SCF-MO-Berechnung der Bindungsabstände in makrocyclischen Bindungssystemen. Tetrahedron 23: 149

74 Ege G, Vogler H (1972a) Zur Konjugation in makrocyclischen Bindungssystemen XX - Charakterordnung, magnetische Suszeptibilitäten und chemische Verschiebungen von Corannulenen. Theor Chim Acta 26: 55

75 Ege G, Vogler H (1972b) Zur Konjugation in makrocyclischen Bindungssystemen XXI - Aromatizität von Corannulenen - Mittlere magnetische Suszeptibilitäten und Exaltationen von Corannulenen. Z Naturforsch 27b: 918

76 Ege E, Vogler H (1975) Zur Konjugation in makrocyclischen Bindungssystemen - V - 1H-chemische Verschiebungen annulenoider Systeme. Tetrahedron 31: 569

77 Elk SB (1980) A Nomenclature for Regular Tessellation and Its Application to Polycyclic Aromatic Hydrocarbons. Match 8: 121

78 Elk SB (1982) Refinement of Systematic Nomenclature for Polybenzenes and Its Extension to Systems of General Arenes. Match 13: 239

79 Elk SB (1985) Formulation of a Canonical Nomenclature for Polybenzenes Using Triangular-Shaped Hexagonal Tessellation Envelopes. Match 17: 235

80 Funhoff DJH, Staab HA (1986) Cyclo[d.e.d.e.e.d.e.d.e.e]Decakisbenzene, A New Cycloarene. Angew Chem Int Ed Engl 25: 742

81 Gaoni Y, Sondheimer F (1964) Conformational Isomers of [14]Annulene. Proc. Chem Soc: 299

82 Gutman I (1982) Topological Properties of Benzenoid Molecules. Bull Soc Chim Beograd 47: 453

83 Gutman I, Cyvin SJ (1988) All-Benzenoid Systems (2) - Topological Properties of Benzenoid Systems - LVII. Match 23: 175

84 Gutman I, Cyvin SJ (1989a) Conjugated Circuits in Benzenoid Hydrocarbons. J Mol Struct (Theochem) 184: 159

85 Gutman I, Cyvin SJ (1989b) Introduction to the Theory of Benzenoid Hydrocarbons. Springer-Verlag, Berlin

86 Gutman I, El-Basil S (1984) Topological Properties of Benzenoid Systems - XXIV - Computing the Sextet Polynomial. Z Naturforsch 39a: 276

87 Gutman I, Milun M, Trinajstić N (1971) Hückel Molecular Orbital Calculation of Aromatic Stabilization of Annulenes. Croat Chem Acta 44: 207

88 Haddon RC, Haddon VR, Jackman LM (1971) Nuclear Magnetic Resonance Spectroscopy of Annulenes. Springer-Verlag, Berlin; Topics in Current Chemistry 16: 103

89 Hall GG (1988) Molecules with Holes. Theor Chim Acta 73: 425

90 Harary F, Harborth H (1976) Extremal Animals. J Combinat Inf & System Sci 1: 1

91 He WC, He WJ (1985) A Novel Nomenclature of Polycyclic Aromatic Hydrocarbons Without Using Graph Centre. Theor Chim Acta 68: 301

92 He WC, He WJ (1986) Generation and Enumeration of Planar Polycyclic Aromatic Hydrocarbons. Tetrahedron 42: 5291

93 He WC, He WJ (1987) One-to-One Correspondence Between Kekulé and Sextet Patterns [in] Graph Theory and Topology in Chemistry (King RB, Rouvray DH, Edit). Elsevier, Amsterdam; Studies in Physical and Theoretical Chemistry 51: 484

94 He WC, He WJ (1990a) Some Topological Properties of Normal Benzenoids and Coronoids. Match 25: 225

95 He WC, He WJ (1990b) Some Topological Properties and Generation of Normal Benzenoids. Match 25: 237

96 He WC, He WJ (1990c) Peak-Valley Path Method on Benzenoid and Coronoid Systems [in] Advances in the Theory of Benzenoid Hydrocarbons (Gutman I, Cyvin SJ, Edit). Springer-Verlag, Berlin; Topics in Current Chemistry 153: 195

97 He WJ, He WC (1986) One-to-One Correspondence Between Kekulé and Sextet Patterns. Theor Chim Acta 70: 43

98 He WJ, He WC (1987) On Kekulé Structure and P-V Path Method [in] Graph Theory and Topology in Chemistry (King RB, Rouvray DH, Edit). Elsevier, Amsterdam; Studies in Physical and Theoretical Chemistry 51: 476

99 He WJ, He WC, Wang QX, Brunvoll J, Cyvin SJ (1988) Supplements to Enumeration of Benzenoid and Coronoid Hydrocarbons. Z Naturforsch 43a: 693

100 Hellwinkel D (1970) Das Corannulen-Konzept. Chemiker-Zeitung 94: 715

101 Herndon WC (1979) Resonance Theory - VI - Bond Orders. J Am Chem Soc 96: 7605

102 Herndon WC, Bruce AJ (1987) Perimeter Codes for Benzenoid Aromatic Hydrocarbons [in] Graph Theory and Topology in Chemistry (King RB, Rouvray DH, Edit). Elsevier, Amsterdam; Studies in Physical and Theoretical Chemistry 51: 491

103 Hess Jr BA, Schaad LJ, Agranat I (1978) The Aromaticity of Annuleno-annulenes. J Am Chem Soc 100: 5268

104 Irngartinger H, Leiserowitz L, Schmidt GMJ (1970) Zur Konjugation in makrocyclischen Bindungssystemen, XVII - Kristall- und Molekülstruk-turen von Hexa-m-phenylen und Penta-m-phenylen. Chem Ber 103: 1132

105 Jenny W, Peter R (1965a) Coronaphene, eine neue Gruppe cyclischer Kohlenwassestoffe. Angew Chem 77: 44

106 Jenny W, Peter R (1965b) Ein zyklischer Kohlenwasserstoff der Benzo-(c)phenanthrenereihe. Chimia 19: 45

107 Kirby EC (1990) Why can so few Benzenoids be Completely Drawn with Clar's Resonant Sextets? An Analysis Using 'Branching Graphs' and 'Coiled-hexagon Code'. J Chem Soc Faraday Trans 86: 447

108 Klarner DA (1965) Some Results Concerning Polyominoes. Fibonacci Quarterly 3: 9

109 Klarner DA (1967) Cell Growth Problems. Can J Math 19: 851

110 Knop JV, Müller WR, Szymanski K, Trinajstić N (1985) Computer Gene-ration of Certain Classes of Molecules. SKTH/Kemija u industriji (Association of Chemists and Technologists of Croatia), Zagreb

111 Knop JV, Müller WR, Szymanski K, Trinajstić N (1986) A Note on the Number of Circulenes. Match 20: 197

112 Knop JV, Müller WR, Szymanski K, Trinajstić N (1990a) Use of Small Computers for Large Calculations - Enumeration of Polyhex Hydrocar-bons. J Chem Inf Comput Sci 30: 159

113 Knop JV, Müller WR, Szymanski K, Trinajstić N (1990b) A Note on the Classification and Enumeration of Coronoid Hydrocarbons. J Mol Struct (Theochem) 205: 361

114 Knop JV, Müller WR, Szymanski K, Trinajstić N (1990c) Enumeration of Planar Polyhex Hydrocarbons. Reports on Molecular Theory: in press

115 Knop JV, Szymanski K, Jeričević Z, Trinajstić N (1983) Computer Enume-ration and Generation of Benzenoid Hydrocarbons and Identification of Bay Regions. J Comput Chem 4: 23

116 Knop JV, Szymanski K, Jeričević Z, Trinajstić N (1984) On the Total Number of Polyhexes. Match 16: 119

117 Knop JV, Szymanski K, Klasinc L, Trinajstić N (1984) Computer Enumera-tion of Substituted Polyhexes. Computers & Chemistry 8: 107

118 Krieger C, Diederich F, Schweitzer D, Staab HA (1979) Molecular Struc-ture and Spectroscopic Properties of Kekulene. Angew Chem Int Ed Engl 18: 699

119   Lunnon WF (1972) Counting Hexagonal and Trigonal Polyominoes [in] Graph Theory and Computing (Read RC, Edit). Academic Press, New York: 87

120   McWeeny R (1951) The Diamagnetic Anisotropy of Large Aromatic Systems - III - Structures with Hexagonal Symmetry. Proc Phys Soc A64: 921

121   Meissner UE, Gensler A, Staab HA (1976) Benzo[14]annulen. Angew Chem 88: 374

122   Meissner UE, Gensler A, Staab HA (1977) Benzo[18]annulen. Tetrahedron Letters: 3

123   Meissner U, Meissner B, Staab HA (1973) Cyclooctadeca[cdefg]phenanthren - ("[18]Annuleno[cdefg]phenanthren"). Angew Chem 85: 957

124   Meyer H, Staab HA (1969) Zur Konjugation in makrocyclischen Bindungssystemen, XIV - Darstellung und Eigenschaften eines ortho-para-verknüpften Nonaphenylens. Liebigs Ann Chem 724: 30

125   Müllen K (1984) Reduction and Oxidation of Annulenes. Chem Rev 84: 603

126   Müller WR, Szymanski K, Knop JV, Nikolic S, Trinajstic N (1989) On Counting Polyhex Hydrocarbons. Croat Chem Acta 62: 481

127   Müller WR, Szymanski K, Knop JV, Nikolic S, Trinajstic N (1990) On the Enumeration and Generation of Polyhex Hydrocarbons. J Comput Chem 11: 223

128   Nikolić S, Trinajstić N, Knop JV, Müller WR, Szymanski K (1990) On the Concept of the Weighted Spanning Tree of Dualist. J Math Chem: in press

129   Ohkami N (1990) Graph-Theoretical Analysis of the Sextet Polynomial - Proof of the Correspondence Between the Sextet Patterns and Kekulé Patterns. J Math Chem 5: 23

130   Ohkami N, Motoyama A, Yamaguchi T, Hosoya H, Gutman I (1981) Graph-Theoretical Analysis of the Clar's Aromatic Sextet - Mathematical Properties of the Set of the Kekulé Patterns and the Sextet Polynomial for Polycyclic Aromatic Hydrocarbons. Tetrahedron 37: 1113

131   Otsubo T, Gray R, Boekelheide V (1978) Bridged [18]Annulenes - 12b,12c,12d,12e,12f,12g-Hexahydrocoronene and Its Mono- and Dibenzo Analogues - Ring-Current Contribution to Chemical Shifts as a Measure of Degree of Aromaticity. J Am Chem Soc 100: 2449

132   Peter R, Jenny W (1966) Höhere, kondensierte Ringsysteme 1.Mitteilung - Untersuchungen in der [10]-Coronaphenereihe - Synthese von Di-[Benzo(c)phenanthren-3,10-Dimethylen]. Helv Chim Acta 49: 2123

133 Polansky OE, Gutman I (1979) Graph-Theoretical Treatment of Aromatic Hydrocarbons IV - Identifizierung zusammenhängender schlichter Graphen als Polyhexes. Match 5: 227

134 Polansky OE, Gutman I (1980) Graph-Theoretical Treatment of Aromatic Hydrocarbons V - The Number of Kekulé Structures in an All-Benzenoid Aromatic Hydrocarbon. Match 8: 269

135 Polansky OE, Rouvray DH (1976a) Graph-Theoretical Treatment of Aromatic Hydrocarbons I - The Formal Graph-Theoretical Description. Match 2: 63

136 Polansky OE, Rouvray DH (1976b) Graph-Theoretical Treatment of Aromatic Hydrocarbons II - The Analysis of All-Benzenoid Systems. Match 2: 91

137 Polansky OE, Rouvray DH (1977) Graph-Theoretical Treatment of Aromatic Hydrocarbons III - Corona-Condensed Systems. Match 3: 97

138 Ramaraj R, Balasubramanian K (1985) Computer Generation of Matching Polynomials of Chemical Graphs and Lattices. J Comput Chem 6: 122

139 Randić M (1975) Graph Theoretical Derivation of Pauling Bond Orders. Croat Chem Acta 47: 71

140 Randić M (1980) Local Properties of Benzenoid Hydrocarbons. Pure & Appl Chem 52: 1587

141 Randić M (1983) On the Role of Kekule Valence Structures. Pure & Appl Chem 55: 347

142 Randić M (1986) A Statistical Approach to Resonance Energies of Large Molecules. Chem Phys Letters 128: 193

143 Randić M, Gimarc BM, Nikolić S, Trinajstić N (1989) On the Aromatic Stability of Helicenic Systems. Gazetta Chim Italiana 119: 1

144 Randić M, Nikolić S, Trinajstić N (1988) Enumeration of Kekulé Structures for Helicenic Systems. Croat Chem Acta 61: 821

145 Randić M, Trinajstić N (1984) Conjugation and Aromaticity of Corannulenes. J Am Chem Soc 106: 4428

146 Sachs H (1984) Perfect Matchings in Hexagonal Systems. Combinatorica 4: 899

147 Schweitzer D, Hausser KH, Vogler H, Diederich F, Staab HA (1982) Electronic Properties of Kekulene. Mol Phys 46: 1141

148 Sondheimer F (1963) Recent Advances in the Chemistry of Large-Ring Conjugated Systems. Pure & Appl Chem 7: 363

149 Sondheimer F (1971) Recent Progress in the Annulene Field. Pure & Appl Chem 28: 331

150 Sondheimer F (1972) The Annulenes. Acc Chem Res 5: 81

151 Sondheimer F (1974) Nonbenzenoid Aromatic (pi)-Electron Systems. Chimia 28: 163

152 Sondheimer F, Gaoni Y (1960) Unsaturated Macrocyclic Compounds - XV - Cyclotetradecaheptaene. J Am Chem Soc 82: 5765

153 Sondheimer F, Wolovsky R (1959) The Synthesis of Cycloöctadecanonaene, a New Aromatic System. Tetrahedron Letters (3): 3

154 Sondheimer F, Wolovsky R (1962) Unsaturated Macrocyclic Compounds - XXI - The Synthesis of a Series of Fully Conjugated Macrocyclic Polyene-polyynes (Dehydro-annulenes) from 1,5-Hexadiyne. J Am Chem Soc 84: 260

155 Sondheimer F, Wolovsky R, Amiel Y (1962) Unsaturated Macrocylic Compounds XXIII - The Synthesis of the Fully Conjugated polyenes Cycloöctadecanonaene ([18]Annulene), Cyclotetracosadodecaene ([24]Annulene), and Cyclotriacontapentadecaene ([30]Annulene). J Am Chem Soc 84: 274

156 Staab HA, Binnig F (1964) Synthese und Eigenschaften von Hexa-m-phenylen. Tetrahedron Letters: 319

157 Staab HA, Binnig F (1967a) Zur Konjugation in makrocyclischen Bindungssystemen, VII - Synthese und Eigenschaften von Hexa-m-phenylen und Octa-m-phenylen. Chem Ber 100: 293

158 Staab HA, Binnig F (1967b) Zur Konjugation in makrocyclischen Bindungssystemen, IX - Penta-m-phenylen und Deca-m-phenylen. Chem Ber 100: 889

159 Staab HA, Bräunling H (1965) Zur Konjugation in makrocyclischen Bindungssystemen II - Synthese und Eigenschaften des 3,6';3',6";3",6-Triphenanthrylens. Tetrahedron Letters: 45

160 Staab HA, Bräunling H, Schneider K (1968) Zur Konjugation in makrocyclischen Bindundssystemen, X - Über 3.6';3'.6";3".6-Triphenanthrylen und verwandte Verbindungen.Chem Ber 101: 879

161 Staab HA, Diederich F (1983) Cycloarenes, A New Class of Aromatic Compounds - I - Synthesis of Kekulene. Chem Ber 116: 3487

162 Staab HA, Diederich F, Čaplar V (1983) Cycloarenes, A New Class of Aromatic Compounds, III - Studies Towards the Synthesis of Cyclo[d.e.-d.e.e.d.e.d.e.e]Decakisbenzene. Liebigs Ann Chem: 2262

163 Staab HA, Diederich F, Krieger C, Schweitzer D (1983) Cycloarenes, A New Class of Aromatic Compounds, II - Molecular Structure and Spectroscopic Properties of Kekulene. Chem Ber 116: 3504

164 Staab HA, Graf F, Doerner K, Nissen A (1971) Zur Konjugation in makrocyclischen Bindungssystemen, XIX - Benzo[12]annulene - Darstellung, Stereochemie und transannulare Reaktionen von Tribenzo[a.e.i]cyclododecenen. Chem Ber 104: 1159

165 Staab HA, Graf F, Junge B (1966) Zur Konjugation in makrocyclischen Bindungssystemen III - Synthese und Eigenschaften von Tribenzocyclododekahexaen. Tetrahedron Letters: 743

166 Staab HA, Meissner UE, Gensler A (1979) Konjugation in makrocyclischen Bindungssystemen, XXIX - Synthese und Eigenschaften von Benzo[18]-annulen. Chem Ber 112; 3907

167 Staab HA, Meissner UE, Meissner B (1976) Zur Konjugation in makrocyclischen Bindungssystemen, XXIII - Phenanthro[cdefg][18]annulen und Dibenzo[ab,de][18]annulen. Chem Ber 109: 3875

168 Staab HA, Meissner UE, Weinacht W, Gensler A (1979) Konjugation in makrocyclischen Bindungssystemen, XXVIII - Benzo-anellierte [14]Annulene. Chem Ber 112: 3895

169 Staab HA, Sauer M (1984) Cycloarene, eine neue Klasse aromatischer Verbindungen, IV - Versuche zur Synthese des Cyclo[d.e.e.d.e.e.d.e.e]-nonakisbenzens und des Cyclo[d.e.d.e.d.e.d.e.d.e]decakisbenzens. Liebigs Ann Chem: 742

170 Staab HA, Wünsche C (1968) Zur Konjugation in makrocyclischen Bindungssystemen, XI - Zur Stabilität aromatischer Kohlenwasserstoffe unter Elektronenbeschuss. Chem Ber 101: 887

171 Stojmenović I, Tošić R, Doroslovački R (1986) Generating and Counting Hexagonal Systems [in] Proceedings of the Sixth Yugoslav Seminar on Graph Theory, Dubrovnik 1985 (Tošić R, Acketa D, Petrović V, Edit). Novi Sad: 189

172 Stollenwerk AH, Kanellakopulos B, Vogler H, Jurić A, Trinajstić N (1983) Magnetic Susceptibilities and Resonance Energies of Annelated [14]- and [18]Annulenes. J Mol Struct (Theochem) 102: 377

173 Thulin B, Wennerström O (1976) Synthesis of [2.2](3,6)Phenanthrenophanediene. Acta Chem Scand B30: 369

174 Tošić R, Doroslovački R, Gutman I (1986) Topological Properties of Benzenoid Systems - XXXVIII - The Boundary Code. Match 19: 219

175 Tošić R, Kovačević M (1988) Generating and Counting Unbranched Catacondensed Benzenoids. J Chem Inf Comput Sci 28: 29

176 Trinajstić N (1983) Chemical Graph Theory, Vols I, II. CRC Press, Boca Raton, Florida

177 Trinajstić N (1990a) On the Classification of Polyhex Hydrocarbons. J Math Chem 5: 171

178 Trinajstić N (1990b) The Role of Graph Theory in Chemistry. Reports on Molecular Theory: in press

179 Trinajstić N, Klein DJ, Randić M (1986) On Some Solved and Unsolved Problems of Chemical Graph Theory. Internat J Quant Chem, Symposium 20: 699

180 Vögtle F, Staab HA (1968) Zur Konjugation in makrocyclischen Bindungssystemen, XII - Versuche zur Darstellung des Cyclo[d.e.d.e.d.e.d.e.-d.e.d.e]Dodecakisbenzens - Eine neue Synthese von 1.2;7.8-Dibenzo-Anthracenen. Chem Ber 101: 2709

181 Vogler H (1979) Calculation of 1H-Chemical Shifts of Kekulene and Similar Compounds. Tetrahedron Letters: 229

182 Vogler H (1980) Zero-Field Splitting Parameters D of Macrocyclic Systems. Croat Chem Acta 53: 667

183 Vogler H (1985) Theoretical Study of Geometries and 1H-Chemical Shifts of Cycloarenes. J Mol Struct (Theochem) 122: 333

184 Vogler H, Ege G (1976) Conjugation in Macrocyclic Bond Systems - VI - 1H-Chemical Shifts of Annelated Annulenes. Tetrahedron 32: 1789

185   Vogler H, Trinajstić N (1988) The Conjugated Circuit Model - On the
     Geometries of Annelated [n]Annulenes. Theor Chim Acta 73: 437

186   Yamamoto K, Harada T, Nakazaki M, Naka T, Kai Y, Harada S, Kasai N
     (1983) Synthesis and Characterization of [7]Circulene. J Am Chem Soc
     105: 7171

S. J. Cyvin, University of Trondheim;
I. Gutman, University of Kragujevac

Kekulé Structures
in Benzenoid Hydrocarbons

1988. XV, 348 pp. (Lecture Notes in Chemistry, Vol. 46)
Softcover DM 74,– ISBN 3-540-18801-0

Contents: Introduction. – Benzenoid Systems: Basic Concepts. –
Kekulé Structures and Their Numbers: General Results. –
Introduction to the Enumeration of Kekulé Structures. –
Non-Kekuléan and Essentially Disconnected Benzenoid
Systems. – Catacondensed Benzenoids. – Annelated Benze-
noids. – Classes of Basic Benzenoids (I). – Classes of Basic
Benzenoids (II): Multiple Zigzag Chain. – Regular Three-,
Four- and Five-Tier Strips. – Classes of Basic Benzenoids (III). –
Classes of Basic Benzenoids (IV): Rectangles. – Regular Six-
Tier Strips and Related Systems. – Deter-
minant Formulas. – Algorithm: A General-
ization. – Pericondensed All-Benzenoids
and Related Classes. – Benzenoids with
Repeated Units. – Distribution of $\underline{K}$, and
Kekulé Structure Statistics. – Bibliography.
– Subject Index.

Springer-Verlag
Berlin
Heidelberg
New York
London
Paris
Tokyo
Hong Kong
Barcelona

I. Gutman, University of Kragujevac;
S. J. Cyvin, University of Trondheim

Advances in the Theory of Benzenoid Hydrocarbons

With contributions by numerous experts

1990. X, 289 pp. 127 figs. 3 tabs. (Topics in Current Chemistry, Vol. 153)
Hardcover DM 178,- ISBN 3-540-51505-4

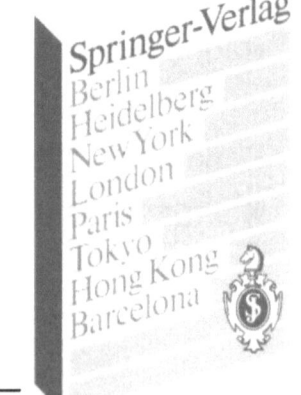

Springer-Verlag
Berlin
Heidelberg
New York
London
Paris
Tokyo
Hong Kong
Barcelona